CREATION COL

MW00769097

VOLCANOES
Earth's Explosive Past

CREATION COLLECTION

VOLCANOES

Earth's Explosive Past

Steven A. Austin, John D. Morris,
Timothy Clarey, Brian Thomas, Jake Hebert,
Frank Sherwin, and James J. S. Johnson

Dallas, Texas
ICR.org

Volcanoes
Earth's Explosive Past

Contributions by Steven A. Austin, Ph.D., John D. Morris, Ph.D., Timothy Clarey, Ph.D., Brian Thomas, Ph.D., Jake Hebert, Ph.D., Frank Sherwin, D.Sc. (Hon.), and James J. S. Johnson, J.D., Th.D.

First Printing: May 2024

Designed by Susan Windsor

Concept by the ICR Communications team: Jayme Durant, Beth Mull, Michael Stamp, Lori Fausak, Renée Dusseau, Bethany Trimble, Rachel Brown, Melissa Marquez, and Frank Sherwin

All Scripture quotations are from the New King James Version.

ISBN (paper): 978-1-957850-60-3
ISBN (ebook): 978-1-957850-62-7
Library of Congress Control Number: 2024936158

Please visit our website for other books and resources: ICR.org

Printed in the United States of America.

Published by Wayfinders Press, an imprint of ICR Publishing Group

TABLE OF CONTENTS

INTRODUCTION

On May 18, 1980, Mount St. Helens spectacularly erupted. The mountain's northern face collapsed, launching pyroclastic flows and a massive landslide that devastated the surrounding area for miles. It's considered the most destructive volcanic event in United States history.

But the eruption also opened a window into Earth's geologic past. For decades, Mount St. Helens has been an outdoor laboratory for studying the effects of catastrophic events. What geologists have found contradicts the conventional account of slow geologic development over millions of years. Instead, evidence shows that a watery catastrophe involving colossal volcanic activity radically transformed Earth's surface just thousands of years ago.

In *Volcanoes: Earth's Explosive Past*, ICR scientists investigate these dynamic geologic features and their role in shaping our planet. Modern volcanoes provide clues to the supervolcanoes that fueled the global Flood and following Ice Age. And the swift recovery of ecosystems after recent eruptions demonstrates how life could have quickly populated the post-Flood world.

Earth itself speaks to the truth of early Genesis history. It says we can trust the accuracy and authority of the Bible in all that it says.

Mount St. Helens eruption on May 18, 1980

1
MOUNT ST. HELENS: A LIVING LABORATORY

Timothy Clarey, Ph.D., and
Frank Sherwin, D.Sc. (Hon.)

Nothing put a damper on uniformitarianism like the Mount St. Helens eruption on May 18, 1980. That explosion still echoes through the halls of the scientific establishment. For nearly 150 years prior to the eruption, strict uniformitarianism reigned supreme in geology. The influence of James Hutton and his concept of deep time had trickled down to even the smallest details. Every geological process was thought to proceed as slowly as those observed today. Erosion and deposition were seen as steady, methodical processes requiring vast amounts of time to make a substantial impact.

In 1980, Mount St. Helens dropped an outdoor laboratory in geologists' laps, forcing them to accept catastrophic events as major contributors to Earth's overall geologic story. Many geologists call this "actualism" as opposed to uniformitarianism.[1] They now acknowledge the evidence that catastrophic events make major impacts on the rock record and that the normal, everyday processes of deposition and erosion contribute very little.

Mount St. Helens has even impacted the science of biology. Flora and fauna recovered very quickly in the devastated area around the volcano. Many centuries aren't required to turn a lifeless terrain into a lush biota filled with life. Plant and animal repopulation is remarkably fast. All of this has implications for understanding the impact and aftermath of the worldwide Flood recorded in Genesis.

Rapid Deposition

The sudden shift in thought was caused by scientists' direct observation of the effects of Mount St. Helens' volcanic activity. Geologists documented that in the six years following the first eruption in 1980, up to 400 feet of new strata had formed at the volcano.[2] These deposits originated from air fall, pyroclastic flows, landslides, and even stream water. The geologists saw that laminated deposits (thin layers) can be produced quickly. Previously, laminated strata were believed to take many years to form, with possibly one layer laid down each year. We now know this assumption is false. One deposit at Mount St. Helens resulted in the creation of a 25-foot-thick, finely laminated unit in a matter of hours![2]

Studies show that rapid deposition is the norm, not the exception. Conventional science has used the slow deposition of sediments like clay and lime mud (micrite) as an argument for an old earth, claiming that all clays form by slowly settling out of stagnant water. People have been indoctrinated with the notion that enormous periods of time are necessary to explain these thick rock layers.

We do see clay settling out of stagnant water today, but the clay-rich rocks we observe didn't form that way. Rocks like shale and mudstone often exhibit fine laminations a few millimeters thick. These layers didn't result from deposition in stagnant water. Empirical evidence demonstrates that laminated clays must be deposited in energetic settings by moving water.[3] The results match the predictions of creation geologists who interpret clay, which forms mudstones and shales, as rapid deposits that occurred during the year-long Flood.[4]

A second finding also has uniformitarian geologists bewildered. Although some lime-rich rocks called carbonates have been interpreted as forming in high-energy settings, carbonate mud has always been thought of as forming in "quiescent ocean settings."[5] But laboratory studies show that micrite is also deposited by moving water. Laminated limestones, like laminated mudstones, aren't only the result of a slow settling process as was previously thought.

According to the authors of a 2013 study, "these experiments demonstrate unequivocally that carbonate muds can also accumulate in energetic settings."[6] They added, "Observations from modern carbonate environments and from the rock record suggest that deposition of carbonate muds by currents could have been common throughout geologic history."

Logs adrift on Spirit Lake, 2012

Rapid Erosion

Mount St. Helens also demonstrated that erosion can be much quicker than previously claimed. The eruption's steam blast, ash flows, and volcanic mud-flows rapidly changed the landscape surrounding the volcano and its waterways. The North Fork of the Toutle River had to carve a new course since the 1980 eruption blocked the original passage with nearly a cubic mile of debris.[2]

After a small subsequent eruption on March 19, 1982, a mudflow from melted snow and ice flowed

A 2019 tour group walks along the log mat at the edge of Spirit Lake. The trees were torn from the northern slope of the lake by a giant wave generated by the massive landslide that initiated the 1980 Mount St. Helens eruption.

down the North Fork of the Toutle River Valley, carving a new canyon up to 140 feet deep.[2] This "Little Grand Canyon" is an approximately 1/40th-scale version of Grand Canyon, demonstrating the rapid scouring power of water. Creation geologists frequently use this as an analogy to explain the rapid formation of the much larger Grand Canyon. Erosion can be fast under the right conditions, and creation geologists believe the global Flood provided ample water to carve canyons and erode mountains in a short amount of time. This was especially true in the receding phase of the Flood, when water energetically washed off and through the soft, sediment-laden landmasses into the newly forming ocean basins during continental uplift.

In fact, it's been well known for decades that even uniformitarian rates of erosion are still so fast that the continents themselves should have been reduced to sea level long ago.[7] A 2011 study confirmed that the average erosion rate of exposed rocks is about 40 feet per million years.[8] This would completely erode most continents in less than 50 million years, and yet they still exist.

Rapid Removal of Vegetation

The 1980 eruption of Mount St. Helens sent the largest landslide ever recorded down the north side of the mountain. Over 3.3 billion cubic yards of rock and ice moving at speeds exceeding 150 mph tore the side of the mountain open, unleashing a devastating steam blast. About 25% of this material ended up in Spirit Lake.[9]

The 680 million cubic yards of material that hit Spirit Lake formed a gigantic tsunami that ripped across the hillsides north of the lake.[9] This water wave sheared off an estimated one million fully grown trees near their bases and transported the logs back to the lake as the water receded. Many of these trees have been found floating upright with the root end down. Following a survey of the floating trees in 1985, it was estimated that more than 19,000 upright logs had settled on the floor of the lake.[2]

Dr. Harold Coffin of the Geoscience Research Institute found that many of the upright trees were randomly spaced, not clumped together, across the bottom of the lake.[2] He noted that many of the trees had settled at various levels in the mud, giving the appearance of deposition at different times.

Creation geologists use Spirit Lake's floating log mat and sunken logs to better understand the likely devastation that occurred during the Flood year. The sunken upright trees are used to explain the numerous polystrate trees often found extending through coal beds. They also help explain petrified forests like the one at Specimen Ridge in Yellowstone National Park. There we find upright trees at different stratigraphic levels that could have formed at the same time during the global Flood.

Forty years later, a massive log mat remained floating around Spirit Lake. However, if this mat had been rapidly buried by subsequent sediments, it's likely these trees would have turned into a coal bed. Coal beds also don't take vast amounts of time to form; they just need the right conditions.

Rapid Recovery of Flora and Fauna

When Noah and his family looked out on the bleak, barren, post-Flood landscape, they must have wondered how long it would remain that way. After a natural disaster, creation scientists observe environmental recovery processes and extrapolate to the worldwide reclamation after the Flood. Mount St. Helens continues to be a scale model of the world God destroyed and reformed as a result of His judgment.

Scripture states that "all the fountains of the great

New growth, Johnston Ridge Observatory

deep were broken up" as the Flood began (Genesis 7:11). This breakup likely included worldwide volcanic activity that continued all over the earth for 150 days (Genesis 7:24; 8:2). Imagine the devastation!

Could there be any recovery after such complete destruction? Surtsey is a good example of a mature and varied landscape that developed in just a few months following the island's volcanic formation in the Atlantic Ocean in 1963.[10] A life scientist who studied it stated in 2008, "Surtsey always provides sur-

Reforestation

Surtsey erupting south of Iceland in 1963

prises....We discover about 20 new species [of life forms] each year."[11] Together, about 60 plant species including mosses, lichens, and an evergreen shrub have been established since Surtsey formed. Its rapidly growing ecosystem is powerful evidence against critics who claim that Earth could not possibly have recovered yet from a worldwide flood that happened only thousands of years ago.

In 2015, research was published[12] regarding "river ecosystems [that] show 'incredible' initial recovery after dam removal" in the western United States.[13] A related article stated:

> During his time conducting the studies in Washington, [ecologist Christopher] Tonra watched reservoir beds that looked like moonscapes return to vibrant, rich habitat and cascades emerge where none had been, at least for the last century. "Watching that happen was just incredible," he said.[13]

The most significant volcanic upheaval in 20th-century America witnessed a similar remarkable renewal of a devastated environment.

When Mount St. Helens erupted in 1980, it destroyed every living thing around it. Gas, ash and rock, heated to over 1000 degrees Fahrenheit, sterilized a 60-kilometer square area, leaving a gray lunar-looking landscape devoid of plants and animals. Within a year, the first plant life had started to return, just as ecologists predicted it would.[14]

In fact, "the recovery of the Mount St. Helens area was 'a wonderful living laboratory' to investigate how ecosystems and species respond to and recover from major disturbances, said Charlie Crisafulli, a research ecologist."[15] Biologists at the University of Washington observed the recovery of two areas covered by violent types of debris flows or mudflows called lahars. They found that "there are striking differences—the forest-surrounded lahar has recovered much faster and has pines and firs atop it, while the more isolated lahar is still mostly covered by grasses, early-stage colonizers."[15]

The Mount St. Helens area has quickly recovered from intense devastation. After only 20 years, biologists noted a rapid recovery of animals and plants on what had been something close to a thermonuclear blast zone. After 40 years, the area was a lush forest. Noah and his family no doubt witnessed the same kind of rapid recovery in the decades following the global Flood.

Conclusion

Mount St. Helens has provided decades of empirical data that support catastrophism and refute strict uniformitarianism. The eruptions have even changed the way conventional scientists view Earth's processes, shifting them to be more accepting of catastrophism. Creation scientists will continue to use Mount St. Helens as a living laboratory to study the devastating effects of events like the global Flood and Earth's rapid recovery in miniature. It is truly a lasting monument to catastrophe.

References

1. Wicander, R. and J. S. Monroe. 2016. *Historical Geology: Evolution of Earth and Life Through Time*, 8th ed. Boston, MA: Cengage Learning.

2. Austin, S. A. 1986. Mt. St. Helens and Catastrophism. *Acts & Facts*. 15 (7).

3. Schieber, J., J. Southard, and K. Thaisen. 2007. Accretion of Mudstone Beds from Migrating Floccule Ripples. *Science*. 318 (5857): 1760–1763.

4. Snelling, A. A. 2009. *Earth's Catastrophic Past*, vol. 2. Dallas, TX: Institute for Creation Research, 493–499.

5. Boggs Jr., S. 2006. *Principles of Sedimentology and Stratigraphy*, 4th ed. Upper Saddle River, NJ: Pearson/Prentice Hall, 159–167.

6. Schieber, J. et al. 2013. Experimental Deposition of Carbonate Mud from Moving Suspensions: Importance of Flocculation and Implications For Modern and Ancient Carbonate Deposition. *Journal of Sedimentary Research*. 83 (11): 1025–1031.

7. Blatt, H., G. Middleton, and R. Murray. 1980. *Origin of Sedimentary Rocks*, 2nd ed. Englewood Cliffs, NJ: Prentice-Hall, Inc.

8. Portenga, E. W. and R. R. Bierman. 2011. Understanding Earth's eroding surface with ^{10}Be. *GSA Today*. 21 (8): 4–10.

9. Morris, J. D. and S. A. Austin. 2003. *Footprints in the Ash: The Explosive Story of Mount St. Helens*. Green Forest, AR: Master Books.

10. Thorarinsson, S. 1967. *Surtsey: The New Island in the North Atlantic*. S. Eysteinsson, trans. New York: Viking Press.

11. An Island Laboratory. 2008. *Science Illustrated*. May/June, 42–47.

12. Tonra, C. M. et al. 2015. The rapid return of marine-derived nutrients to a freshwater food web following dam removal. *Biological Conservation*. 192: 130.

13. River ecosystems show 'incredible' recovery after dam removal. Ohio State News. Posted on new.osu.edu December 28, 2015, accessed March 5, 2020.

14. Mount St. Helens Recovery Slowed By Caterpillar. University of Maryland, College Park. *ScienceDaily.* Posted on sciencedaily.com November 16, 2005, accessed March 5, 2020.

15. Thompson, A. Mount St. Helens Still Recovering 30 Years Later. *Live Science.* Posted on livescience.com May 17, 2010, accessed March 5, 2020.

2
HOW MOUNT ST. HELENS
REFUTES EVOLUTION

Brian Thomas, Ph.D.

When the ICR Discovery Center was under construction, ICR was pleased to host Gary Bates, CEO of Creation Ministries International-USA, for a private tour.[1] While we watched busy workers add greenery to the Mount St. Helens model, Gary told of a time years ago when evidence from ICR's research of the 1980s eruptions in Washington State led him to switch from evolution to creation.

Within days of becoming a Christian, Gary began struggling to fit evolution with the Bible. His struggle ended when he saw geologist Dr. Steve Austin present ICR research on Mount St. Helens. By the end of that hour, he had learned enough geology to refute fish-to-fishermen evolution. What does a volcanic eruption have to do with belief in evolution? Gary explained his logic.

First, he learned that Mount St. Helens released the energy of 20 million tons of TNT on the morning of May 18, 1980. The blast pulverized rock and ejected tons of steam-infused ash and sediments throughout

Mount St. Helens eruption on May 18, 1980

that day. Over three billion cubic yards of avalanche material slid down.[2] Eventually, the muddy wreckage settled. At this point, Gary expected that the debris had settled haphazardly.

But Dr. Austin spoke of a 1982 mudflow that carved deep channels through the thick 1980 deposits, exposing mud made solid in just two years. It shocked Gary to learn that the initial explosion formed distinct layers. Giant cross-beds and fine, flat layers both formed fast. He realized that it doesn't take a million years to make layers. You just need plenty of fast-moving water.

The Mount St. Helens events only needed hours and months to form the same features found in sed-

Gary Bates, center, describes his experience the first time he heard an explanation of the Mount St. Helens eruptions in light of the Genesis creation narrative.

imentary rocks around the world. Common features include:

- Sharp, flat contacts between layers
- Larger particles toward the bottom of a rock bed
- Cross-beds
- Steep-walled canyons
- Drainage systems
- Material moved far away before becoming part of new rock
- De-limbed, sorted, and reburied logs
- Volcanic ash mixed with mud and hardened into rock

As soon as Gary learned that rock layers can form fast, he knew Noah's Flood could have produced enormous rock stacks in one year. We don't need long ages to solidify sediments. But what about the age-dating schemes that supposedly prove conventional ages?

Within six years of the eruption, a new lava dome in the crater atop Mount St. Helens had hardened. Standard radioisotope methods pointed to an isotopic age of around 350,000 years for the 10-year-old rock.[3] Gary learned then that the highly regarded radioisotope dating methods are broken.[4] This made Noah's recent Flood that much more sensible.

He reasoned that if rock layers formed fast, then the earth could be only thousands of years old—that's not enough time for fish to evolve into people.

Unimaginably long timespans form the backdrop for all evolutionary speculations, but the world's rocks show evidence of rapid deposits. Erase the time and you erase evolution. I respect Gary for following the evidence where it led—to biblical creation and thus to the trustworthiness of the God who inspired His Holy Word.

References

1. Based near Atlanta, Georgia, Creation Ministries International-USA promotes biblical creation.

2. Morris, J. and S. A. Austin. 2003. *Footprints in the Ash*. Green River, AR: Master Books, 25.

3. Ibid, 67.

4. See Dr. Vernon Cupps' book *Rethinking Radiometric Dating: Evidence for a Young Earth from a Nuclear Physicist*, published by the Institute for Creation Research.

Mount St. Helens, North Fork Toutle River Channel

3
REMEMBERING MOUNT ST. HELENS

Brian Thomas, Ph.D.

The volcano's main 1980 eruption filled in an entire valley with hundreds of feet of sediment. Another smaller eruption event deposited more material on top of that, and then a third deposition occurred in 1982. Later, a catastrophic flood of snowmelt water and muddy debris tore a gash through those fresh deposits, revealing sharp, flat contacts between each earlier deposit. It also showed that fast-flowing currents can lay down multiple layers thinner than a finger's width.

Dr. John Morris examines the canyon formed by a 1982 lahar, revealing layered deposits from 1980 and 1982 eruptions

Mount St. Helens revealed to the world that both thick and thin layering can happen fast. Millions of years aren't needed to form sedimentary rock or stratigraphic layering.

Sedimentary layers hundreds of feet thick formed

Stream cutting through the pyroclastic flow layers, Mount St. Helens Volcanic National Monument, Washington

within hours during the eruption itself and then hardened into rock soon after the water drained from them. Could other layered sedimentary rocks in Earth's crust have formed rapidly?

The mountain also provided a clear reason to distrust radioisotope dating. Geologist Dr. Steve Austin sampled new rock from atop the mountain that formed in 1986. If the K-Ar radioisotope method really works, then it should have revealed the rock's true age of only 10 years. Instead, three rock ages ranged from 340,000 to 2,800,000 years.[1] What other rocks from around the world have been dated incorrectly by following those same questionable age-dating protocols?

Because of the Mount St. Helens eruption, sci-

Lava dome in caldera in 2005

entists know that sedimentary rock layers and steep-walled canyons can form in only hours. The documented evidence is undeniable. It also exposes flaws in radioisotope dating. Because of this, catastrophism has become readily accepted in the geological sciences. How many other Earth surface features will scientists recognize as having formed suddenly or much more recently than previously recognized?

Reference

1. Austin, S. A. 1996. Excess Argon within Mineral Concentrates from the New Dacite Lava Dome at Mount St. Helens Volcano. *Creation Ex Nihilo Technical Journal.* 10 (3): 335–343.

4
BIOLOGICAL BOUNCEBACK AT MOUNT ST. HELENS

Brian Thomas, Ph.D.

Early one bright Sunday morning, Mount St. Helens looked as picturesque as it had for hundreds of years. Suddenly, an earthquake shook loose the north side of the mountain. This uncorked a torrent of heat, ash, and steam that torched trees, pulverized rock, emptied and elevated Spirit Lake, and blasted debris across the nearby landscape. New land formed, mudflows wrecked bridges, and sediment clogged waterways far downstream. That was over 40 years ago, but the eruption still teaches valuable lessons today.[1]

United States Geological Survey research hydrologist Jon Major has published several reviews in remembrance of the radical eruption of Mount St. Helens in Washington State. He described rapid recolonization of the utterly devastated landscape.

The vicious May 18, 1980, eruption was supposed to have wiped all life from the north slope, but some survived the blast. Major wrote in a review for the American Geophysical Union publication *Eos*, "Even some of the most heavily affected landscapes

30

were not as sterile as initially assumed."[2] He wrote in *Science*, "Notably, remnants of the pre-eruption biota—biological legacies—that persisted even in what appeared to be a lifeless landscape critically affected ecological recovery."[3]

Before the eruption, conventional ecologists thought living things would slowly creep back into the ash and mud from surrounding areas. They thought it could take a century or more for biology to brave the barren blast zone. Evolutionary thinking influenced those preconceived ideas. Nowadays, ecologists see the rapid recolonization at Mount St. Helens

Recent vegetation thrives near Mount St. Helens. This photo was taken in June 2017, 37 years after the 1980 eruption.

1983

2014

Two views showing plant development between 1983 and 2014 at upper Smith Creek, an area affected by the pyroclastic blast. The regrowth reflects individual survivors as well as colonizing plants.

as a good reason to recalibrate their old ways of thinking. And these updated expectations for how fast life can colonize mesh well with creation thinking.

Genesis says that God created creatures to multiply and fill the earth. If true, then He would have equipped them with the proper tools to do just that. No wonder keystone species, which support other species in the ecosystem, soon took root, even in nutrient-poor soil.

Major cited research showing that alder trees and lupines began to grow in the mud first. These pioneering plants form symbiotic relationships with microbes that partner with their root tissues, pulling vital nitrogen from the air and plugging it into the soil.

Pocket gophers' burrowing brought that newly nutritive soil to the surface. So in far less time than expected, ground that was near keystone species had

Smith Creek regrowth over 14 years

enough nutrition to support other plants. Today, tall alder and other trees sprinkle the grassy landscape that not too long ago looked quite like the moon's surface.

Major wrote in *Eos*, "The importance of biological legacies in promoting recovery emerged as an epiphany."[2] This biological recovery research has influenced land management philosophy. Major wrote of "using variable-retention harvesting, rather than clear-cutting."[2] Creation thinking fits this idea, too. Instead of wiping out every single tree when harvesting wood, we should wisely leave keystone species behind. God equipped them with the tools to quickly re-establish a new forest that harbors habitats, builds biodiversity, and makes more wood that much faster.

In contrast to the pre-1980s evolution-inspired

thinking that each organism struggles for its own survival, the big ecological lesson from Mount St. Helens is that plants and animals often work together to help one another grow and thrive. Creatures partner to pioneer new lands just like Genesis says they were made to do.

References

1. Clarey, T. and F. Sherwin. 2020. Mount St. Helens, Living Laboratory for 40 Years. *Acts & Facts*. 49 (5): 10–13.

2. Major, J. J. Lessons from a Post-Eruption Landscape. *Eos*. Posted on eos.org April 24, 2020.

3. Major, J. J. 2020. Mount St. Helens at 40. *Science*. 368 (6492): 704–705.

5
WHY DOES ICR STUDY THE MOUNT ST. HELENS ERUPTION?

John D. Morris, Ph.D.

Ever since the May 18, 1980, eruption of Mount St. Helens, ICR has made it a focus of intense research. From it we have learned a great deal about the origin of rocks and geologic features and the processes needed to form them.

In general, ICR holds that most of Earth's rocks were formed rapidly during the great Flood of Noah's day, not over the supposed millions of years of geologic history. But here's the problem: geologists like to study modern rocks and the processes that form them and infer past circumstances. But Noah's Flood was a totally unique event, unlike any in our experience. Those geologists who assume uniformity in history thus seem to have an advantage. But the rocks really do appear to have been formed by dramatic processes operating at rates, scales, and intensities far beyond those we experience. Only modern, local catastrophes, such as the eruption of Mount St. Helens, can give us a glimpse into Earth's geologic power, particularly as we expand our thinking onto the worldwide scale of Noah's Flood. Thus, the Mount St. Helens catastrophe becomes a scale model for the great Flood.

Keep in mind that most of the damage done by the eruption was water-related. Mount St. Helens had been glacier-covered, and when it got hot, water raced down the mountain as a mighty flood, eroding soil, rocks, trees, and everything else in its path, eventually redepositing them at the foot of the mountain. Volcanic episodes added to the fury. When the eruption calmed, up to 600 feet of sediments had been deposited, full of plant and animal remains. Now the

Crater Lake National Park, Oregon. When this now-dormant volcano erupted several thousand years ago, its ash emission volume was about 70 times that of the Mount St. Helens eruption.

sediments have hardened into sedimentary rock and the dead things have fossilized. Furthermore, wood is petrifying. Peat (the precursor to coal) has formed. A deep canyon has been gouged out. Many features that geologists are taught take long ages to form were seen to happen rapidly. Igneous rocks formed since 1980 yield radioisotope dates of millions of years but are obviously *much* younger.

A catchy slogan helps illustrate this. To form geologic features, it either takes a little bit of water and a long time, or a lot of water and a short time. Even though we didn't witness the Flood, we do see modern catastrophes, and they rapidly accomplish things the Flood did on a grander scale. In a short, biblically compatible time scale, such a flood can account for the features we see on Earth, features that many geologists mistake for evidence of great age. Earth doesn't really look old, it looks flooded.

Aerial view of Yellowstone's Grand Prismatic Spring

6
SUPERVOLCANOES AND THE MOUNT ST. HELENS ERUPTION

Steven A. Austin, Ph.D.

The eruption of Mount St. Helens volcano marked a turning point in geologists' understanding of volcanic processes. That eruption became the geologic event of the 20th century. Mount St. Helens was not the largest volcanic episode of the last century, but it became the most informative. May 18, 1980, was the explosive day that convinced many geologists that catastrophic geologic processes need to be reintroduced into geologic thinking. The lesson became obvious: uniformitarianism and Darwinism were thwarting the practice of geology.[1]

Mount St. Helens erupted one-quarter cubic mile of magma through a nozzle that day. After that, geologists coined a new word to describe colossal volcanic events—supervolcano.[2] They were convinced that enormous chambers of magma could be erupted along fissures as well as through nozzles. Modern nozzle eruptions (such as Krakatoa in 1883 and Mount St. Helens in 1980) usually eject less than three cubic miles of ash.

Such nozzle eruptions are trivial compared to the ancient fissure events that created supervolcanoes, in which more than 240 cubic miles of magma were erupted. These colossal volcanoes were over a thousand times larger than Mount St. Helens. For example, Ice Age supervolcanoes, such as Long Valley of California and Yellowstone in Wyoming, exploded just after the Flood. Unlike nozzle eruptions, these were ring-fissure eruptions. A vertical crack opened to vent ash, then the crack unzipped in a vast circle and finally collapsed catastrophically to form a circular or elliptical depression called a caldera.

Crater Lake in Oregon is just a very small example of a caldera from a circular ring fissure. Yellowstone and Long Valley are some of Earth's largest calderas. In northwestern Italy, a 16-mile-thick succession of rock was turned sideways by the late Flood catastrophic plate collision between Europe and Africa.[3] The sideways succession of rock displays the understructure of an eight-mile-wide caldera, allowing us to visualize the plumbing system that sustained some of the earth's greatest volcanic explosions.

Even bigger supervolcanoes erupted earlier during the Flood. These larger supervolcanoes occurred as linear-fissure-arrays eruptions. These are parallel fractures that do not turn to form an ellipse or circle. For example, the Independence Dike Swarm in southern California was caused when catastrophic plate tectonics opened numerous parallel fissures a hundred miles long. The array of linear fissures extends southward from east-central California to Baja California.[4] This eruption deposited the Brushy Basin

View of Krakatoa during the Earlier Stage of the Eruption.
from a Photograph taken on Sunday the 27th of May, 1883.

Krakatoa erupted in 1883. It killed around 60,000 people.

Member of the Morrison Formation ("Upper Jurassic") of the Rocky Mountain and Great Basin regions, which today contain more than 4,000 cubic miles of ash.[5] Within the Brushy Basin ash are sandstone beds that represent the gigantic mudflows that swept up and buried dinosaurs.

Not just volcanic ash but also colossal lava flows issued from linear fissure arrays on both the continents and on the ocean floor. The Columbia River Basalts of eastern Washington and Oregon were erupted from parallel feeder dikes in southeastern Washington. Seafloor lava flows called the Nikolai Greenstone are up to seven miles thick and have been docked by tectonic processes to form southern Alaska.

The eruption of Mount St. Helens provides an opportunity to pause and reflect on the supervolcanoes of the past. We see the fury of Flood volcanic activity and the declining power of post-Flood volcanism.[6] Mount St. Helens is almost trivial when compared to previous explosive events. Yet, the eruption of May 18, 1980, has opened a window to the turbulent volcanic world of the past, providing further confirmation of the Bible's depiction of a young earth that was catastrophically shaped by the global tectonics of the Flood.

References

1. Morris, J. D. and S. A. Austin. 2003. *Footprints in the Ash*. Green Forest, AR: Master Books.

2. The word "supervolcano" was coined by the BBC documentary *Supervolcanoes*, first shown February 3, 2000.

3. Quick, J. E. et al. 2009. Magmatic plumbing of a large Permian caldera exposed to a depth of 25 km. *Geology*. 37 (7): 603–606.

4. Austin, S. A. and W. A. Hoesch. 2006. Do Volcanoes Come in Super-Size? *Acts & Facts.* 35 (8).

5. Hoesch, W. A. and S. A. Austin. 2004. Dinosaur National Monument: Jurassic Park or Jurassic Jumble? *Acts & Facts.* 33 (4).

6. Austin, S. A. 1998. The Declining Power of Post-Flood Volcanoes. *Acts & Facts.* 27 (8).

Mount St. Helens

7
VOLCANOES OF THE PAST
John D. Morris, Ph.D.

During the great Flood of Noah's day, God unleashed His great power, exercising His righteous judgment on the wicked and violent people that rejected Him. He promised not only to judge sinful man but also the earth (Genesis 6:13). All of Adam's dominion (Genesis 1:26) came under the sin penalty because of his choice to disobey God. By the time of Noah's day, rebellion had increased so much that God finally enacted His just penalty for sin (Romans 6:23). He sent the worldwide Flood to punish the wicked, purge the entire planet, and start over with the descendants of righteous Noah.

The Flood primarily involved hydraulic processes, with rainfall pummeling the earth for 150 days (Genesis 7:11–8:4). The "fountains of the great deep" also broke open, spewing onto the surface of the earth huge volumes of water, magma, and whatever else was beneath the earth's crust (Genesis 7:11). "Fountains" suggest tectonic activity as well, both on land and under water.

Through an understanding of today's volcanic

Thingvellir, Iceland, where two tectonic plates dramatically meet on the earth's surface

eruptions, we can better comprehend those of the past. However, the rock record suggests that yesterday's volcanoes were evidently supervolcanoes, accomplishing geologic work hardly comparable to those we currently observe.

If we plot the volume of ash and lava extruded by volcanoes throughout history—comparing Vesuvius (AD 79) and Krakatoa (1883) to more recent volcanoes such as Mount St. Helens (1980) and Pinatubo (1991)—we conclude that the earth processes are quieting down. And if we plot the materials blown out by volcanoes that erupted during the great Flood and soon thereafter (inferred only from the materials left behind), then

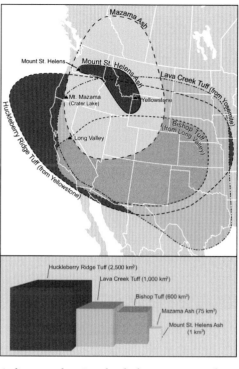

A diagram showing the declining magnitude of volcanic eruptions. Those occurring during and soon after the great Flood were immense compared to more recent ones.

we see an exponential decline in the power of Earth's volcanoes over time. This implies that Flood volcanoes were many times greater than those recently witnessed.

Earth underwent a complete tectonic restructuring during the great Flood, with supervolcanoes, mega-earthquakes, supercurrents of flowing water and mud, and hypercanes. All of these exhibit a similar exponential decline in intensity. Thankfully, we do

1991 Pinatubo pyroclastic flow

not witness comparable events, but by studying the impact of the great Flood, we can begin to understand how much God hates sin. Likewise, we can surmise the nature of the coming judgment, when the earth will pass away and be replaced by a new earth (2 Peter 3:10–13).

Reference

1. Diagram modified from Morris, J. and S. Austin. 2003. *Footprints in the Ash.* Green Forest, AR: Master Books, 12.

8
YELLOWSTONE NATIONAL PARK: A FLOOD SUPERVOLCANO

Timothy Clarey, Ph.D., and Brian Thomas, Ph.D.

Yellowstone became the world's first national park in 1872. Geologist Ferdinand Hayden led an expedition the year before through much of what became the 2.2-million-acre park, and his report helped convince the U.S. Congress to set aside the land for that purpose. It occupies the northwestern corner of Wyoming and spills into Idaho and Montana.

The park is unique for its travertine flows, mud pots, hot springs, and canyons. It houses over 10,000 thermal features and over half of the world's geysers.[1] Its otherworldly landscapes and raw wildlife have been sufficient cause to protect the region for generations. This park's features fit well with biblical history.

Volcanic Beginnings

Yellowstone is one of about 12 supervolcanoes in the world. Thick layers of volcanic ejecta from huge eruptions cover most of the area and beyond. The middle of the park is a volcanic plateau that lies about 7,300 feet above sea level. Yellowstone Lake rests within the plateau, and mountain peaks taller than 11,000 feet surround it.

The central plateau has three overlapping calderas, or collapsed volcanoes. Each collapse occurred after a major eruption. The largest eruption occurred first, and the smallest happened last. This trend reveals diminishing power over time instead of the steady processes that uniformitarians imagine. That scenario fits what we would expect as Earth recovered from the tumultuous Flood about 4,500 years ago.[2]

Geologists noticed a line of older calderas that extends from Yellowstone to Oregon. Conventional scientists consider this trail to have formed from a mantle hotspot as Earth's crust slowly moved across it over millions of years.[3] But millions-of-years age assignments rely on an assumed evolutionary timeline, not directly on data.

The Institute for Creation Research recognizes that the evidence that Earth's tectonic plates moved much faster during the Flood year continues to build.[4] It is likely that relatively rapid movement of North America over a mantle hotspot more quickly formed the volcanic trail, perhaps within months. Catastrophic conditions slowed dramatically after the Flood. Today, we see little plate motion or volcanism at Yellowstone.

The magma source at Yellowstone is high in silica (quartz-rich). This resulted in especially violent eruptions. In contrast, basalt magmas are lower in silica and flow more gently, like those erupting in

The Yellowstone portion of ICR's 2021 science expedition. Left to right: Media Director Michael Hansen, Events Director Joel Kautt, and Research Scientists Drs. Timothy Clarey and Frank Sherwin at the Grand Canyon of the Yellowstone River overlook.

Hawaii. Silica-rich continental crust likely altered the deep mantle magmas that rose to the surface at Yellowstone. This unique magma chemistry also helps explain the park's geysers and hot springs.

Geysers

Geysers are hot springs that emit jets of hot water and steam. They need an active shallow heat source to superheat groundwater into steam. Yellowstone provides that with magma detected just three miles below the surface.[5]

Geysers also need a fairly watertight plumbing system.[6] The silica-rich magma provides that, too. Silica dissolves in circulating hot groundwater. It then drops out of solution (precipitates) onto the fracture surfaces and conduits to form a hard deposit called sinter. This deposit seals water in and makes constrictions that allow geysers to build up pressure.[6]

Underground water expands as it warms until spaces inside the conduits become compressed, like a pressure cooker. Steam and fluids rise and in places reach Earth's surface at nearly 200°F, sometimes jetting into the sky.[6] Creation scientists speculate how the geysers might illustrate the eruptions through great fractures that took place around the planet when the fountains of the great deep burst open at the start of the Flood (Genesis 7:11).

Excelsior Geyser Crater

Old Faithful

Yellowstone's most popular geyser, Old Faithful, may not produce the tallest eruption—although its 130-feet-plus average is impressive—but, as its name implies, it's one of the most predictable. This regularity stems from its steady source of underground water.

Old Faithful

But Old Faithful does change. Several decades ago, it averaged less time between eruptions. Local tremors have since lengthened that time.[7] Geysers like Old Faithful point to the role that water played in the formation of Earth and then its reformation during the Flood (2 Peter 3:5–6).

Rainbow Pools

Rainbow-colored hot springs like Grand Prismatic Spring may present the most striking of Yellowstone's natural wonders. Different microbial mats form rainbow-like rings when viewed from above. Ferdinand Hayden stated in 1871, "Nothing ever conceived by human art could equal the peculiar vividness and delicacy of color of these remarkable prismatic springs."[8]

Hot water rising from underground cools as it flows from the center of the pools. Heat-loving (thermophilic) microbes called *Synechoccus* use unique biochemistry to pioneer the perimeter of the 370-foot diameter of the Grand Prismatic Spring. They live there at about 150°F (65.6°C).[9]

These cyanobacteria harvest sunlight to obtain energy. In the summer, they deploy yellow pigments to protect their vital DNA from the sun's UV rays. Remarkably, they use those same pigments to reassign that would-be-damaging light energy toward their green pigments (chlorophylls) for photosynthesis.

Outside the pool's yellow ring grows a mixture of microbes that together look orange. An even wider red ring lies outside of that. The wider variety of mi-

Grand Prismatic Spring

crobes here also use their pigments for protection and light harvesting, but each variation harvests a different part of the light spectrum available for photosynthesis. As the microbes share the light, they convert and store its energy into sugars.

These marvelous microbes form strings, filaments, and mats as they trade resources. Surface dwellers give their sugars to microbes beneath them and receive other nutrients in return. None of this works without precise microengineering, for which an engineer, the Lord Jesus, deserves full credit.

Mammoth Hot Springs

Hot springs can bring up dissolved lime, sulfur, and even mud. North of Yellowstone's main caldera, Mammoth Hot Springs conveys two tons of dissolved lime from depth per day.[5]

The water here likely gets its heat from Earth's thermal gradient (increasing temperature with depth) far beneath the surface. The complex at Mammoth Hot Springs covers nearly a square mile and deposits more lime than any other spring in the world.[6]

As the water cools at the surface, lime precipitates to form travertine, a form of limestone, in a complex of terraces. Travertine deposits can grow fast under the right conditions. Extrapolating today's deposition rate into the past shows that this hot spring complex has been active for only thousands of years.

Conclusion

Yellowstone is a beautiful and awe-inspiring reminder of the global Flood. Calderas and thousands of feet of volcanic rock bear witness to great catastrophes that occurred during the Flood year. The magma that powers active geysers has not yet fully cooled, which is what would be expected from the relatively short time since the Flood.

Christ Jesus' creative genius explains the microbes that collaborate to thrive in extremely hot waters. The hot springs display part of the Lord's provision of the hydrologic cycle that moves water around our planet. Although the Flood destroyed the world that then was, it left behind beauty and wonders that point us to Him.

References

1. Tweit, S. J. 1999. Yellowstone. In *America's Spectacular National Parks*. L. B. O'Connor and D. Levy, eds. Los Angeles, CA: Perpetua Press, 76–79.

2. Austin, S. A. 1998. The Declining Power of Post-Flood Volcanoes. *Acts & Facts*. 27 (8).

3. A hotspot is thought to form from a near-stationary deep mantle plume of high heat capable of producing a melt through the overriding crust.

4. Clarey, T. Plate Subduction Beneath China Verifies Rapid Subduction. *Creation Science Update*. Posted on ICR.org December 23, 2020, accessed February 12, 2022; Clarey, T. 2020. *Carved in Stone: Geological Evidence of the Worldwide Flood*. Dallas, TX: Institute for Creation Research.

5. How big is the magma chamber under Yellowstone? U.S. Geological Survey FAQ. Posted on usgs.gov, accessed February 17, 2022.

6. Hacker, D. and D. Foster. 2018. Yellowstone National Park: Northwest Wyoming, Eastern Idaho, Southern Montana. In *The Geology of National Parks*, 7th ed. D. Hacker, D. Foster, and A. G. Harris, eds. Dubuque, IA: Kendall-Hunt, 765–791.

7. Milstein, M. Old Faithful slows, but grows. *Billings Gazette*. First captured online January 17, 2001. Retrieved from web.archive.org January 11, 2022.

8. Theurer, J. Colors of Curiosity: Yellowstone's Microbes. *Yellowstone Quarterly*. Posted on yellowstone.org April 16, 2019, accessed January 13, 2022.

9. Geiling, N. The Science Behind Yellowstone's Rainbow Hot Spring. *Smithsonian Magazine*. Posted on smithsonianmag.com May 7, 2014, updated May 12, 2016, accessed January 14, 2022.

9
YELLOWSTONE NATIONAL PARK: CANYONS AND CATASTROPHE

Timothy Clarey, Ph.D., and Brian Thomas, Ph.D.

About three million visitors tour Yellowstone National Park's 3,440 square miles each year.[1] Most come to see natural wonders like Old Faithful and Mammoth Hot Springs,[2] but this park packs in much

more. Yellowstone has its own grand canyon, waterfalls, and even upright petrified trees. Its surface features look young and testify to the catastrophic forces that shaped them.

Grand Canyon of the Yellowstone

A deep canyon with three waterfalls cuts through the park's eastern spur. This Grand Canyon of the Yellowstone flows for 20 miles and reaches

Lower Falls of Yellowstone

1,200 feet deep. Before an ancient catastrophe carved the canyon, hot waters moved upward through gray and red volcanic rocks. This hydrothermal action turned them yellow and orange. These yellow stones line the canyon's walls, inspired the park's name, and overlook places like Inspiration Point. When did these geologic events occur?

Water flows from Yellowstone Lake through the Grand Canyon of the Yellowstone over stunning waterfalls. The first cliff forms Upper Falls, dropping a modest 109 feet. Next, the water plummets 308 feet down Lower Falls. Tower Falls' 132-foot drop comes last at the northern end of the canyon.

Conventional geologists believe an Ice Age glacier dammed up a huge lake where Yellowstone Lake is today. When water burst through melted ice, it quickly carved through the rock layers.[3] However, they also believe this river canyon started to form much earlier—about a half million years ago.[4]

Roaring water can erode even hard volcanic rock in a short time. It forms tiny vacuum bubbles that implode with great force in a process called cavitation, which can demolish rock or concrete.[5] After over a half million years, these three waterfalls should have smooth, gentle slopes instead of the steep-sided falls that command visitors' attention today. With Yellowstone's landscape looking only a few thousand years old, it fits well with an Ice Age right after Noah's Flood.[6]

Fossil Tree Trunks and Catastrophe

The northeastern part of the park has petrified

tree trunks embedded in volcanic rock debris. A hike up Specimen Ridge reveals some of them. About 20 different layers in the vicinity contain tree fossils, many standing vertically. Sycamore, walnut, chestnut, oak, maple, redwood, and magnolias were fossilized here.[4] Conventional geologists claim these trees grew in place and that over 20 separate eruptions covered as many separate forests over eons.[4] If so, then where are the roots and branches? Instead, some process stripped, sorted, and reburied these tree trunks.

Lessons from the 1980 Mount St. Helens eruption offer a more fitting model. That historic blast snapped off a forest of trees at their roots, tore off their branches, and then dumped their trunks into Spirit Lake. Some became waterlogged and sank, thickest end down. They landed upright in the volcanic debris at the lake bottom.[7] An even grander catastrophe must have razed, stripped, and interred the many petrified trees at Yellowstone.

The Flood's world-destroying forces had the energy to do this. Patterns of rock layers globally and around Specimen Ridge suggest these trees sank soon after the Flood crested on Day 150 (Genesis 7:24). Floodwaters likely ripped them from the high hills. As floodwaters flowed off Earth's surface, they jostled the logs in a moving debris mat. Some tree trunks settled vertically, and some sank faster than others. All the while, Yellowstone's eruptions trapped the trunks in wet volcanic debris and ash. The 20 different layers would have quickly formed, one right after the other.

It's a tough hike to Specimen Ridge, but visitors can save their legs by taking a turnout to see Petrified Tree. This big tree grew when Noah was alive around 4,500 years ago. It was torn free by the waves, and the Flood buried it. Later, groundwater deposited minerals inside the wood to petrify it.[8]

Yellowstone's striking canyons and waterfalls showcase a recently carved landscape, and upright petrified tree trunks make stories of multiple forests sound improbable. The park's clear evidence of the catastrophic forces that shaped it testify to the global Flood recorded in Genesis and the Ice Age that followed.

References

1. Tweit, S. J. 1999. Yellowstone. In *America's Spectacular National Parks*. L. B. O'Connor and D. Levy, eds. Los Angeles, CA: Perpetua Press, 76–79.

2. Clarey, T. and B. Thomas. 2022. Yellowstone National Park, Part 1: A Flood Supervolcano. *Acts & Facts*. 51 (3): 10–13.

3. Clarey, T. 2017. Minuscule Erosion Points to Hawaii's Youth. *Acts & Facts*. 46 (1): 9.

4. Hacker, D. and D. Foster. 2018. Yellowstone National Park: Northwest Wyoming, Eastern Idaho, Southern Montana. In *The Geology of National Parks*, 7th ed. D. Hacker, D. Foster, and A. G. Harris, eds. Dubuque, IA: Kendall-Hunt, 765–791.

5. Morris, J. 2011. The Channeled Scablands. *Acts & Facts*. 40 (10): 15.

6. Hebert, J. 2018. The Bible Best Explains the Ice Age. *Acts & Facts*. 47 (11): 10–13.

7. Austin, S. A. 1986. Mt. St. Helens and Catastrophism. *Acts & Facts*. 15 (7).

8. Thomas, B. and T. Clarey. 2021. The Painted Desert: Fossils in Flooded Mud Flats. *Acts & Facts*. 50 (4): 16–19.

10
YELLOWSTONE SUPERVOLCANO UNLIKELY TO BLOW

Timothy Clarey, Ph.D.

For several years, conventional scientists have been predicting a possible supervolcano eruption at Yellowstone National Park. In 2020, the London *Daily Mail* reported that the Norris Geyser Basin in Yellowstone National Park, an area about the size of Chicago, has risen about six inches from 2013 to 2015.[1]

This brought on more speculation and fear of another supervolcano eruption that would potentially impact most of the United States. This area of Yellowstone has already shown two similar rises in ground elevation in the recent past, but they have subsided each time.[1] Modeling of geophysical data suggested that magma came within nine miles of the surface below the geyser basin between 1996 and 2004.[1] As stated in the *Daily Mail*,

> When magma intrudes the crust it cools, crystallizes, and releases gases that had been dissolved in the melt. Gas escape lowers pressure in the magma, causing the surface to subside...But rising gases can become trapped

under an impermeable layer of rock, causing the kind of rapid uplift seen at Norris from late 2013 until the [magnitude] 4.9 earthquake in March 2014.[1]

Dan Dzurisin, a member of the science team monitoring the elevation of the land over Norris Geyser Basin, said, "It seems likely the quake created microfractures that allowed gases to escape upward again, resulting in subsidence that ended in 2015."[1] This ended the second active episode.

Dzurisin further reported,

The third uplift episode from 2016 to 2018 suggests rising gases became trapped again, this time at a slightly shallower depth. For the first time, we've been able to track an entire episode of magma intrusion, degassing, and

Yellowstone River in Hayden Valley—the northeastern part of Yellowstone Caldera

gas ascent to the near-surface.[1]

How does all of this factor into the big picture of Yellowstone? And should we be concerned about a major volcanic eruption?

Today we see an ocean ridge system coming north up the Gulf of California (along the Baja Peninsula) and a second ocean ridge system off the west coast of Washington and Oregon, creating the subduction zone there and the Cascade Mountains. These two ridge systems are connected through the San Andreas Fault system. The former section of the ridge system between them has been subducted beneath North America.

A subducted ridge system tends to be more buoyant and exerts pressure upward on the base of the continent. This might provide a cause for the base-

ment-cored mountain ranges so far inland from the subduction zone that existed along the West Coast. Kelleher and McCann first noticed this off the west coast of South America where ridges have been subducted nearly horizontally beneath the continent for hundreds of kilometers.[2]

A subducted ridge system so far inland beneath North America could also explain the trail of the Yellowstone hotspot from Oregon to northwest Wyoming in the Cenozoic. This line of volcanism parallels the northern edge of the ridge system that was subducted beneath North America. The highest point of the ridge system may have funneled magma along the crest of the subducted ridge like the peak of a tent, releasing vast quantities of magma to "burn" a hotspot trail through the overlying crust. In similar fashion, the Yellowstone supervolcano eruptions could also be a consequence of the magma pooling beneath the subducted ridge. But keep in mind, these eruptions only occurred in the receding phase of the Genesis Flood.

Not even all conventional scientists are concerned about a major eruption at Yellowstone. Of course, they believe the last supervolcano eruption occurred about 600,000 years ago.[1] "For those in the know, like you, that's awesome—not alarming," Dzurisin explained.[1]

One thing we can be assured of: God is in control. We do not need to fear another supervolcano eruption at Yellowstone. These past eruptions were part of the global Flood year, about 4,500 years ago. It should be no surprise that magma is still close to the surface there, as evidenced by the geysers and geophysical data. A small eruption could occur, but not a massive supervolcanic eruption that would destroy much of our nation. These eruptions were part of a special judgment as recorded in the book of Genesis. We do not need to live in fear of that today.

References

1. Liberatore, S. Huge chunk of Yellowstone National Park 'breathing' in and out due to trapped magma. *Daily Mail*. Posted on dailymail.co.uk March 20, 2020, accessed March 25, 2020.

2. Kelleher, J. and W. McCann. 1976. Buoyant zones, great earthquakes, and unstable boundaries of subduction. *Journal of Geophysical Research*. 81 (26): 4885–4896.

Mauna Loa eruption in November 2022

11
WORLD'S LARGEST VOLCANO FOUND HIDING UNDER THE OCEAN

Timothy Clarey, Ph.D.

A study published in *Earth and Planetary Science Letters* reveals that Pūhāhonu volcano is the world's largest by volume and is the hottest.[1] Found almost 700 miles northwest of Hawaii, Pūhāhonu volcano is almost completely submerged beneath the Pacific Ocean.[2] Only two small, rocky remnants stick up to about 170 feet above the surface, exposing only six acres.[1]

Prior to this discovery, Mauna Loa volcano on the Big Island of Hawaii was thought to be the largest volcano on Earth.[1] Both Pūhāhonu and Mauna Loa are shield volcanoes. These volcanoes form by massive outpourings of a low-viscosity (flows easily), basalt-rich lava, resulting in a broad, fairly flat volcano that resembles a Greco-Roman shield.

The Hawaiian Islands are at the terminal end of a long chain of volcanoes extending over 3,600 miles across the Pacific Ocean. Most of these volcanoes are now below sea level and are either flat-topped guyots or seamounts. Pūhāhonu volcano is one of the nearly completely submerged volcanoes within this chain. Most scientists think this long line of volcanoes formed from a hotspot in the mantle underneath the ocean plate. As the Pacific plate moved over the hotspot toward the northwest, newer and newer volcanoes formed to the southeast, extending the chain. The Big Island now sits over the active hotspot. All islands and seamounts to the west are progressively older and are essentially inactive.

In 2014, sonar data was collected over Pūhāhonu that was analyzed by Michael Garcia and colleagues at the University of Hawai'i at Mānoa, Honolulu. They found that Pūhāhonu volcano is nearly twice the mass of Mauna Loa, having a volume of about 35,500 cubic miles.[1] They also studied rock samples from the volcano and determined it had the hottest magma source at about 1,700°C, which was even hotter than the Deccan Trap volcanism in India.[1]

The scientists then tried to figure out why this volcano was so big compared to the volcanoes and seamounts around it in the chain. Garcia and his fellow scientist wrote, "We considered four testable mechanisms to increase magma production, including 1) thinner lithosphere, 2) slower propagation rate, 3) more fertile source, and 4) hotter mantle."[1]

They ruled out the first three possibilities and found the fourth one seemed to best fit the data, concluding:

A hotter mantle remains the best mechanism to produce the large magma volumes and is consistent with the high forsteritic olivine phenocryst [crystals] compositions (up to 91.8%) and the calculated high percent of melting (24%). Thus, the gargantuan size of Pūhāhonu reflects its high melting temperature, the highest reported for any Cenozoic basalt.[1]

No one is sure why the magma was hotter at the time that Pūhāhonu volcano sat over the hotspot beneath the crust. Garcia and his team suggested it may have been caused by a "solitary wave" that had formed briefly in the mantle. They explained, "Solitary waves are stable ascending blobs of buoyant, viscous [thick] mantle created by perturbations in the flux entering the conduit." In other words, they speculate there was a higher and hotter flow of magma coming from deep in the mantle, but they are still unsure of a cause.

Creation geologists believe the volcanic ridge system that includes Pūhāhonu volcano and major portions of the islands of Hawaii all formed rapidly

The Gardner Pinnacles—the tip of Pūhāhonu

during the Flood year. As the newly formed Pacific plate was being created along the East Pacific Rise, the plate was moving to the northwest at speeds of several yards per second.[3] This catastrophic plate motion slid the oceanic plate quickly over the nearly stationary hotspot, leaving the resulting trail of volcanoes behind.

There must have been a tremendous outpouring of mantle magma to produce this long ridge of volcanoes so quickly. During the Flood year, runaway subduction and the rapid formation of crust at ocean ridges must have caused flow in the mantle and volcanism like never before.[3] And we see evidence of this globally, with volcanic activity peaking late in the Flood.[3]

As the activity of the Flood year lessened, the plates also slowed down to a crawl, making the massive Mauna Loa volcano at the end of the chain. The reasons why the mid-chain Pūhāhonu volcano attained such an anomalously large volume is still not fully understood. A hotter mantle may just be part of it. But such massive outpourings of lava seem much more understandable in the context of a global catastrophe like the Flood.

References

1. Garcia, M. O. et al. 2020. Pūhāhonu: Earth's biggest and hottest shield volcano. *Earth and Planetary Science Letters*. 542 (116296).

2. Perkins, S. World's biggest volcano is barely visible. *ScienceMag*. Posted on sciencemag.com May 12, 2020, accessed May 18, 2020.

3. Clarey, T. 2020. *Carved in Stone: Geological Evidence of the Worldwide Flood*. Dallas, TX: Institute for Creation Research.

12
DEEP-SEA VOLCANO GIVES GLIMPSE OF FLOOD ERUPTIONS

Timothy Clarey, Ph.D.

A team of scientists from Australia and the U.S. studied the ejecta from a subsea volcano, gaining new insights into how magma can explode to the surface from deep underwater.[1] This discovery also gives important insight into volcanic activity during the Flood year when many volcanoes originated while still underwater.

In 2012, a relatively unnoticed eruption occurred beneath the ocean at Havre volcano. The volcano is about 500 miles northeast of the north island of New Zealand. Its summit is about 3,000 feet below sea level.

"The common theory is that underwater eruptions, particularly in deep water such as at Havre, cannot be explosive and instead make lava flows on the seafloor," lead author Joseph Knafelc told *Science Daily*.[2]

"The problem is that it was an underwater eruption that had to push up through nearly 1 km [3,000 feet] of ocean. The only way it can do this is if the eruption was very powerful and able to punch through the ocean water and produce an eruption column in the air," Knafelc added.[2]

The team, publishing in *Communications Earth & Environment*, found that the minerals in the ejecta suggest a "short-lived but powerful explosive eruption phase penetrated the water column [3,000 feet] allowing hot pyroclasts to oxidise in air."[1]

Knafelc and his colleagues found that the eruption created a powerful jet that shielded the hot material from the cooling effects of the water column until it reached the atmosphere. They concluded that their results challenge the known depth limits for underwater eruptions.

An explosive eruption column could get hot pumice into the atmosphere in as little as a few seconds. This was a very powerful eruption. The problem is that previous studies had not recognised or downplayed the explosive

West Mata volcano

potential of submarine eruptions even in very deep water and thus the hazards posed by submarine eruptions.[2]

Flood geologists are not as surprised by these findings. We suspect that many volcanoes erupted while still under the cover of the Flood waters. Many of the volcanoes around the Pacific Ring of Fire originated underwater and were equally as explosive as Havre. Even the supervolcano at Yellowstone had its beginnings underwater while the Flood was beginning to recede and the Ice Age began.[3]

The eruption of Havre volcano gives us a glimpse into the catastrophic activity that occurred all over the earth during the global Flood. Deep-sea volcanoes were exploding everywhere. They didn't merely produce lava flows on the seafloor, as conventional scientists have previously believed. Instead, their powerful eruptions of pyroclastic material blasted to the surface and spread ash and cinders for many miles in all directions and contributed to the destruction of the pre-Flood world. They also were major contributors to the post-Flood Ice Age.[4]

References

1. Knafelc, J. et al. 2022. Havre 2012 pink pumice is evidence of a short-lived, deep-sea, magnetite nanoliter-driven explosive eruption. *Communications Earth & Environment*. 3 (1).

2. Pink pumice key to revealing explosive power of underwater volcanic eruptions. Queensland University of Technology. Posted on sciencedaily. com February 8, 2022, accessed March 3, 2022.

3. Clarey, T. 2020. *Carved in Stone: Geological Evidence of the Worldwide Flood*. Dallas, TX: Institute for Creation Research, 312–377.

4. Hebert, J. 2013. Was There an Ice Age? *Acts & Facts*. 42 (12): 20.

13
CRATER OF DIAMONDS STATE PARK
AND THE ORIGIN OF DIAMONDS

Brian Thomas, Ph.D.

At Crater of Diamonds State Park in western Arkansas, families dig diamonds for fun while more serious sifters seek sensational paydays. Countless brides have wondered where the pretty diamonds on their rings came from. A visit to Crater of Diamonds reminds us of two key research results that refute the old ages assigned to diamonds. These results favor the Bible's record of a more recent creation and Flood.

An Ancient Volcano

Diamonds adorn the Arkansas state flag and license plates in honor of Crater of Diamonds State Park. The park is a rare place for the public to mine diamonds. The experience draws a diverse crowd. Serious searchers bring wagons toting buckets of gravelly soil to the park's water troughs for wet sifting, while children bring their fingers to poke around the ground.

Despite the crater in the name, first impressions reveal no distinct crater shape at ground level. Rather, volcanic rocks called lamproite mixed in the soil

Prospectors at Crater of Diamonds State Park in Murfreesboro, Arkansas

show that you're walking on the mouth of an ancient volcano. Park signage compares the ancient volcanic blast cloud to one of the largest ever recorded—the 1883 explosion of Krakatoa in Indonesia (misidentified on the park's sign as happening in 1848).

In a moment, mantle material from miles below shot up through Earth's crust. When it neared the surface, exploding steam blasted fresh lava rock into a vast volcanic ash plume. If this erupted while Cretaceous sediments had already dropped to the bottom of the Flood's surging waters, then it happened near the middle of the Flood year.[1] Figure 1 shows the thickness of sediments that landed on southern Arkansas during the fifth of six major pulses of Flood sedimentation, each pulse called a megasequence.

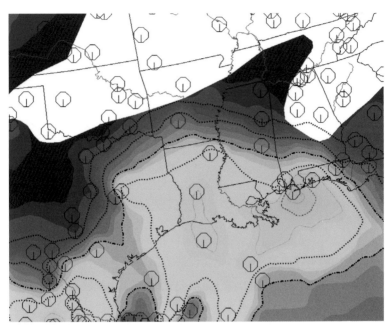

Figure 1. *Isopach (thickness) map of the sedimentary rocks composing the Zuni Megasequence (Jurassic and Cretaceous systems) for Arkansas and surrounding states. The volcano at Crater of Diamonds State Park erupted near the top edge of these sediments, shown in blue. Greener color indicates thicker deposits, and bluer color indicates thinner deposits. Circles show drill core data collection areas.*

If our Flood model is accurate, then the ash plume could have erupted right into floodwaters, not directly into air.

Back then, volcanoes gushed as continents collided. During this fifth megasequence, months into the Flood year, water finally overtopped the highest hills that then existed (Genesis 7:19). As volcanic eruption rates slowed after the Flood, so did diamond deliveries from the deep. Those who appreciate their diamond rings can thank God for bringing up gems through this judgment.

Two Research Results

Does evidence from Earth—or even better, directly from diamonds—confirm our Flood model?

Various park signs repeat the common assertion that diamonds formed billions of years ago, but two key research results bury that idea. The first came from the ICR Radioisotopes and the Age of the Earth (RATE) project.[2] One branch of RATE research involved measuring radiocarbon (radioactive carbon) in diamonds. If Earth formed thousands of years ago, then Earth's diamonds might still retain some radioactive carbon atoms. In contrast, billions-of-years bars carbon-containing material deemed older than 100,000 theoretical years from having any radiocarbon.[3]

Radiocarbon labs use carbonaceous earth materials like natural gas, coal, marble, and graphite as instrument blanks. Workers assume they are too old to have any radiocarbon, but labs consistently reveal more radiocarbon in the blanks than contamination can reasonably account for. And before the RATE project, nobody had tested diamonds.

We express the levels of radiocarbon using pMC. This refers to the percentage of radiocarbon found in an ancient sample compared to the radiocarbon content in a standard modern sample. In general, the older the material, the smaller the percentage. Based on the measured decay rate, samples older than 100,000 theoretical years should show zero percent radiocarbon after subtracting known contamination sources like radiocarbon atoms in the lab air.

In 2005, Dr. John Baumgardner described the

Sample ID	pMC	±
NMBclr1	0.31	0.02
NMBclr2	0.17	0.02
NMBclr3	0.13	0.03
NMByel1	0.09	0.02
Letlhakane-3	0.07	0.02
NMBrn2	0.07	0.02
Letlhakane-1	0.04	0.03
NMByel2	0.04	0.02
Orapa-F	0.03	0.03
Kankan	0.03	0.03
Kimberley-1	0.02	0.03
Orapa-A	0.01	0.03
AMS sensitivity	0.002	

Table 1. *Twelve RATE-tested diamonds showed levels of radiocarbon above the AMS (accelerator mass spectrometer) theoretical sensitivity limit.[4] It appears that short-lived radiocarbon is intrinsic to these diamonds. pMC is percent modern carbon, uncalibrated.*

Lab ID	pmC	±
UCIAMS-12677	0.018	0.001
UCIAMS-12678	0.017	0.002
UCIAMS-12676	0.017	0.001
UCIAMS-12679	0.016	0.001
UCIAMS-12674	0.015	0.001
UCIAMS-12675	0.015	0.001
UCIAMS-15445	0.021	0.003
UCIAMS-15444	0.018	0.002
UCIAMS-15443	0.015	0.002
UCIAMS-15446	0.013	0.002
UCIAMS-15447	0.011	0.002
UCIAMS-9638	0.008	0.001
UCIAMS-9640	0.006	0.001
UCIAMS-9639	0.005	0.001

Table 2. *Six of nine diamonds show levels of radiocarbon far above the instrument's detection limit. The top six measurements were taken from one diamond. UCIAMS refers to the University of California Irvine Accelerator Mass Spectrometer. Results from Taylor and Southon, 2007.[5]*

RATE team's stunning results, shown here in Table 1.[4] All 12 African diamond specimens had radiocarbon in them! With all this radioactive carbon, how could these diamonds be even one million years old, let alone billions? No known underground process could generate the measured levels of radiocarbon.

Conventional experts, including the late R. E. Taylor, decided to measure their own diamonds. They found radiocarbon in theirs, too. Table 2 summarizes their results, published in 2007.[5]

These kinds of results so irked Taylor that he wrote a 10-page paper to try to explain them away.[6] He and his co-authors argued that since diamonds are billions of years old, any radiocarbon they measured must result from contamination. That's poor reasoning based on the data.[7] These results deserve to be investigated, not buried beneath bias.

A visit to Crater of Diamonds State Park, or a look at any diamond ring, reminds us of the short-lived radiocarbon that two research projects found inside diamonds. Both sets of results confirm the Bible's report that Earth is only thousands of years old.

References

1. Clarey, T. 2020. *Carved in Stone: Geological Evidence of the Worldwide Flood.* Dallas, TX: Institute for Creation Research, 285.

2. RATE stands for Radioisotopes and the Age of the Earth. The multifaceted project's predictions were published in 2000, and their spectacular fulfillments were published in 2005. All documents are available online at ICR.org/rate.

3. Given the relatively short half-life of 5,730 years for radiocarbon's decay into nitrogen, none would remain after 100,000 years.

4. Baumgardner, J. 2005. Carbon-14 Evidence for a Recent Global Flood and a Young Earth. In *Radioisotopes and the Age of the Earth: Results of a Young-Earth Creationist Research Initiative.* L. Vardiman et al., eds. San Diego, CA: Institute for Creation Research and Chino Valley, AZ: Creation Research Society, 587–630.

5. Taylor, R. E. and J. R. Southon. 2007. Use of natural diamonds to monitor ^{14}C AMS instrument backgrounds. *Nuclear Instruments and Methods in Physics Research B.* 259: 282–287.

6. Taylor, R. E., J. R. Southon, and G. M. Santos. 2018. Misunderstandings Concerning the Significance of AMS Background [14]C Measurements. *Radiocarbon.* 60 (3): 727–749.

7. Cupps, V. R. and B. Thomas. 2019. Deep Time Philosophy Impacts Radiocarbon Measurements. *Creation Research Society Quarterly.* 55 (4): 212–222.

View of the Middle East from space

14
INTENSE ICE AGE VOLCANISM FITS BIBLICAL MODEL

Frank Sherwin, D.Sc. (Hon.)

Evidence continues to accumulate that the volcanic activity that likely contributed to the Genesis Flood caused an Ice Age lasting hundreds of years.[1]

Creation scientists maintain there are two significant climate factors needed for an Ice Age. First, warm oceans are needed to increase evaporation, which ultimately generates extra rain near the equator and heavy snowfall in higher latitudes. Such conditions would build ice sheets atop the continents.

Second, Earth's atmosphere must contain enough very small airborne particles and droplets called aerosols to reflect sunlight and keep the ice from melting during warm summer months. These aerosols would have come from the explosive volcanic eruptions, which are known to be capable of causing cooler summers.[1]

Eighty-five of the volcanic eruptions identified by the researchers were large global eruptions. Sixty-nine of these are estimated to be larger than the 1815 eruption of Mount Tambora in Indonesia—the largest volcanic eruption in recorded human history. So much sulfuric acid was ejected into the stratosphere by the Tambora eruption that it blocked sunlight and caused global cooling in the years that followed.[1]

Such cooling following the Flood would have allowed ice to accumulate over the decades. In light of all this, it is unsurprising to creation scientists that a 2022 news story stated, "Ice cores drilled in Antarctica and Greenland have revealed gigantic volcanic eruptions during the last ice age."[2,3]

Interestingly, the Bible suggests that the Middle East once received more snow and ice than it does today, so a post-Flood Ice Age does fit what we find in Scripture. For example, the book of Job, which is chronologically the oldest book of the Bible except for the first part of Genesis, records how the climate during the time was cold and icy:

> From the chamber of the south comes the whirlwind, And cold from the scattering winds of the north. By the breath of God ice is given, And the broad waters are frozen.[4]

Although the thick Ice Age ice sheets did not extend all the way to the Middle East, the increased Ice Age precipitation may have made cold weather more common in those areas nearer the equator.

At this point, volcanic cooling is well known. Is there a reason mainstream paleoclimate scientists can't make better use of it to explain an Ice Age? Yes, there is. They believe historical volcanic eruptions were separated by many thousands (and even millions) of years, so any cooling they caused would have been so diluted as to be inconsequential. It is the Bible's short timescale that makes the straightforward Ice Age explanation possible.

The biblical Ice Age model requires intense Ice Age volcanism, and these deep ice cores confirm that such volcanism was indeed present.

More evaporation, warmer winters, more intense storms, and colder summers: The result? An Ice Age that would last until the oceans gave up their excess heat, the volcanism lessened, and vegetation was re-established.

References

1. Hebert, J. 2018. The Bible Best Explains the Ice Age. *Acts & Facts.* 47 (11): 10–13.

2. Ancient ice reveals scores of gigantic volcanic eruptions. *Phys.org.* Posted on phys.org March 16, 2022, accessed March 17, 2022.

3. Actually, creationists would argue there was only one Ice Age, since evidence for other supposed Ice Ages is extremely weak.[1] Indeed, evidence for the mainstream Ice Age theory is also very weak, as shown by original ICR research. See Hebert, J. 2020. Physics Today Article Ignores Monster Milankovitch Problem. *Creation Science Update.* Posted on ICR.org May 24, 2020, accessed March 17, 2022.

4. Job 37:9–10.

Clouds in the stratosphere

15
VOLCANOES, GEOENGINEERING, AND THE POST-FLOOD ICE AGE

Jake Hebert, Ph.D.

A study in 2020 suggested that sulfur dioxide (SO_2) injected into Earth's stratosphere could fight global warming with minimum adverse effects.[1,2]

At ICR, we generally think alarmism over global warming or climate change is unwarranted. We aren't necessarily endorsing this study's proposal, but we can use it to shed light on what really caused the Ice Age.

An Ice Age requires cooler summers. Cooler summers allow some, or most, snow and ice to survive the summer months. The following winter, still more snow and ice would be deposited. If these cooler summers persist for many years, thick ice sheets develop, resulting in an Ice Age.

Scientists have long known that large, explosive volcanic eruptions can cause cooler summers. The year 1816 has been called "the year without a summer" because of its much colder summer temperatures, the result of the large Mt. Tambora eruption in 1815.[3] A 1988 study of explosive volcanic eruptions

The Mt. Tambora caldera

in the late 20th century showed that they caused summer cooling,[4] and this has been confirmed by more recent studies:

> *Warmer than average winters and cooler than average summers* over continental Northern Hemisphere areas have been documented and modeled after several eruptions, including Pinatubo..., and this appears to be part of the normal Northern Hemisphere response after volcanic aerosol events.[5,6]

Explosive volcanic eruptions can belch large amounts of sulfur dioxide into the atmosphere. Atmospheric chemical reactions involving this SO_2 form tiny droplets of sulfuric acid (H_2SO_4) that can remain in the stratosphere for years. These droplets reflect sunlight back into space, resulting in cooler summers.

Since an Ice Age requires cooler summers, explosive volcanic eruptions would seem to be an obvious candidate for this cooling. Yet such eruptions play, at best, only a secondary role in the dominant conventional Ice Age model. The most popular theory (Milankovitch theory) holds that the timing of Ice Ages is controlled by slow, gradual changes in Earth's orbital and rotational motions. Yet, this theory has many problems, and the evidence for the theory (hypothesis, really) is weak at best and non-existent at worst.[7–9] It should also be noted in passing that the assumption that this theory is correct is one of the main arguments for catastrophic climate change, which climate alarmists themselves have acknowledged.[10,11]

So why don't conventional models make better use of volcanic cooling to explain an Ice Age? Both creation and uniformitarian scientists think Earth has experienced a large amount of past volcanic activity. But because uniformitarian scientists insist that these volcanic eruptions were separated by millions of years, any resulting summer cooling is so spread out and "diluted" over vast ages that it really wouldn't do a whole lot. Remember, these cooler summers need to persist for many years in order to get an Ice Age.

However, the Genesis Flood gives us the clues we need to solve this mystery. The hardened volcanic rocks that are evidence of cataclysmic past volcanism are intermingled with water-deposited sedimentary rocks that formed as a result of the Flood. This means that these volcanic eruptions occurred during the year-long Flood (mostly toward the end) and continued afterward for many years as the earth slowly recovered from the catastrophe. These volcanic eruptions, many of which were large and explosive, provided the needed cooling for an Ice Age.[12] ICR's Dr. Timothy Clarey has found that volcanic deposits increase greatly late in the global rock record, peaking in the Cenozoic, when the Flood was receding and just prior to the onset of the Ice Age.[13]

Moreover, creation scientists think catastrophic plate tectonics played a major role in the Flood.[14] This implies very rapid formation of a completely new seafloor, and the heat from this rapidly-extruded lava would have greatly warmed the world's oceans. This would have resulted in much more evaporation, which in turn would have provided the extra moisture needed for heavy snowfall, a second Ice Age requirement.[15]

This is just another example of how, contrary to what some Christians seem to think, the Genesis record is not a problem for which embarrassed Christians need to sheepishly apologize. It is the key to solving mysteries of Earth history that have stumped conventional scientists for literally hundreds of years. Recent biblical creation isn't the problem—it's the solution!

That scientists are considering using a man-made version of volcanic cooling to fight perceived global warming is a reminder that the Bible provides a much better framework for understanding Earth history than conventional models.

References

1. University College London Press Release. The right dose of geoengineering could reduce climate change risks, study says. *Phys.org*. Posted on phys.org March 19, 2020, accessed March 27, 2020.

2. Irvine, P. J. and D. W. Keith. Halving warming with stratospheric aerosol geoengineering moderates policy-relevant climate hazards. *Environmental Research Letters*. 15 (4).

3. Mount Tambora. *Encyclopaedia Britannica*. Posted on britannica.com, accessed March 27, 2020.

4. Bradley, R. S. 1988. The Explosive Volcanic Eruption Signal in Northern Hemisphere Continental Temperature Records. *Climatic Change*. 12 (3): 221–243.

5. Self, S. et al. 1996. The Atmospheric Impact of the 1991 Mount Pinatubo Eruption. In *Fire and Mud: Eruptions and Lahars of Mount Pinatubo, Philippines*. C. G. Newhall and R. S. Punongbayan, eds. Quezon City, Phil, Seattle, WA, and London, UK: Philippine Institute of Volcanology and Seismology and University of Washington Press. Emphasis added.

6. Based on this statement, one might wonder if warmer winters are a problem for the biblical Ice Age model. No, because it is, surprisingly, *colder* winters that are a problem when explaining the Ice Age. Colder air has much less moisture than warm air, which results in *less* snow, not more! In fact, mild winters are expected in the biblical Ice Age model.

7. Oard, M. J. 2007. Astronomical troubles for the astronomical hypothesis of ice ages. *Journal of Creation*. 21 (3): 19–23.

8. Hebert, J. 'Big Science' Celebrates Invalid Milankovitch Paper. *Creation Science Update*. Posted on ICR December 26, 2016, accessed March 25, 2020.

9. Hebert, J. More Problems with Iconic Milankovitch Paper. *Creation Science Update*. Posted on ICR.org July 13, 2018, accessed March 25, 2020.

10. Hebert, J. 2019. Climate Alarmism and the Age of the Earth. *Acts & Facts*. 48 (4): 11–14.

11. Hebert, J. 2019. *The Climate Change Conflict: Keeping Cool Over Global Warming*. Dallas, TX: Institute for Creation Research.

12. Clarey, T. 2019. Subduction Was Essential for the Ice Age. *Acts & Facts*. 48 (3): 9.

13. Clarey, T. 2020. *Carved in Stone: Geological Evidence of the Worldwide Flood*. Dallas, TX: Institute for Creation Research, 349–352.

14. Clarey, T. 2016. Embracing Catastrophic Plate Tectonics. *Acts & Facts*. 45 (5): 8–11.

15. Hebert, J. 2018. The Bible Best Explains the Ice Age. *Acts & Facts*. 47 (11): 10–15.

2022 Tonga eruption

16
THE TONGA VOLCANO ERUPTION AND THE ICE AGE

Jake Hebert, Ph.D.

On January 15, 2022, an underwater volcano in the Pacific Ocean's Kingdom of Tonga erupted with the energy of hundreds of Hiroshima-size atom bombs.[1] Both the resulting column of ash and the shock front that rippled away from the eruption could be seen from satellites in space.[1,2]

Explosive volcanic eruptions can cause cooler temperatures. Global temperatures dropped about 0.6°C after the Philippines' Mt. Pinatubo injected 15 million tons of sulfur dioxide (SO_2) into the atmosphere during its 1991 eruption.[3] Although the Tonga eruption is quite impressive, it is estimated to have generated a much smaller amount of SO_2, only about 0.4 million tons. So, any cooling from the Tonga eruption would be much less pronounced.[4]

The Tonga and Pinatubo eruptions are reminders that explosive volcanic eruptions are one of the key factors needed to explain the Ice Age. An Ice Age requires cooler summers so that winter ice and snow will not completely melt during the summer months.

As more snow and ice are deposited during subsequent winters, thick ice sheets can grow over time. In addition to cooler summers, heavy snowfall is also needed, because sunlight can still melt light snowfall, even when temperatures are below freezing.

The Genesis Flood provided both warmer oceans and cooler summers. Most creation scientists think the pre-Flood ocean floor was subducted into the mantle during the Flood, and a new seafloor was formed during the year of the Flood.[5] The hot lava that rapidly formed the new seafloor greatly warmed the world's oceans, dramatically increasing evaporation rates for many centuries after the Flood. This high moisture content resulted in heavy post-Flood precipitation, including snowfall at high latitudes and on mountaintops.

Explosive volcanic eruptions during the Flood and intermittent explosive volcanic eruptions in the centuries following resulted in cooler summers, especially in the Northern Hemisphere.[6] These cooler summers prevented winter snow and ice from melting, allowing the formation of thick ice sheets.[7]

The Tonga eruption is what volcanologists call a Surtseyan eruption, an explosive eruption that takes place in shallow water. They are named that because such an eruption formed the island of Surtsey off the coast of Iceland in 1963.

The island of Surtsey has long been of interest to creation scientists because some Bible skeptics have argued that there is no way that the earth could have recovered ecologically if the Genesis Flood was as devastating as creationists claim it is. Yet just 45 years after Surtsey was formed from cooling lava, researchers found that insects, birds, and various species of plants had already begun colonizing it.[8]

The conventional Ice Age theory posits that the timing of Ice Ages is controlled by slow changes in Earth's orbital and rotational motions. Despite its widespread acceptance, the theory has major problems, and evidence for the theory is shaky at best, even by conventional reckoning.[9] Volcanic cooling and warm oceans are a much better explanation, and the Bible tells us of an event that would provide those conditions: the Genesis Flood.

References

1. Specktor, B. Tonga eruption equivalent to 'hundreds of Hiroshima bomb,' NASA says. *LiveScience*. Posted on livescience.com January 25, 2022, accessed January 27, 2022.

2. Andrews, R. G. Tonga shock wave created tsunamis in two different oceans. *Science*. Posted on science.org January 25, 2022, accessed January 27, 2022.

3. Global Effects of Mount Pinatubo. 2001. NASA Earth Observatory.

4. Bhan, V. Tongan eruption's Sulphur dioxide could lead to cooler winter. 1 News. Posted on 1news.co.nz January 18, 2022, accessed January 27, 2022.

5. Clarey, T. 2016. Embracing Catastrophic Plate Tectonics. *Acts & Facts*. 45 (5): 8–11.

6. Bradley, R. S. 1988. The Explosive Volcanic Eruption Signal in Northern Hemisphere Continental Temperature Records. *Climatic Change*. 12 (3): 221–243.

7. Hebert, J. 2018. The Bible Best Explains the Ice Age. *Acts & Facts*. 47 (11): 10–13.

8. Sherwin, F. Surtsey: A Young-earth Laboratory. *Creation Science Update*. Posted on ICR.org May 28, 2008, accessed January 27, 2022.

9. Hebert, J. Physics Today Article Ignores Monster Milankovitch Problem. *Creation Science Update*. Posted on ICR.org May 24, 2020, accessed January 27, 2022.

17
SUBDUCTION WAS ESSENTIAL FOR THE ICE AGE

Timothy Clarey, Ph.D.

Creation meteorologist Michael Oard has written extensively about what it takes to make an Ice Age. The first requirement is much warmer oceans than we have today, which would provide the extra evaporation needed for heavy winter snowfall. The second requirement is cooler summers that allow snow to build up from year to year and eventually transform into thick ice sheets.[1]

But what would warm the oceans? And what could cause cooler summers for many years in a row? Catastrophic plate tectonics provides the answer for both warmer oceans and cooler summers. Such tectonic activity appears to be the mechanism God used to implement the global Flood.[2,3] During the Flood, plates rapidly subducted into Earth's mantle and formed hot, new seafloor at the ocean ridges. The result was a peak in ocean heating and volcanism at the same time.

If an explosive volcanic eruption is large enough, its aerosols can cool Earth by blocking out sunlight.

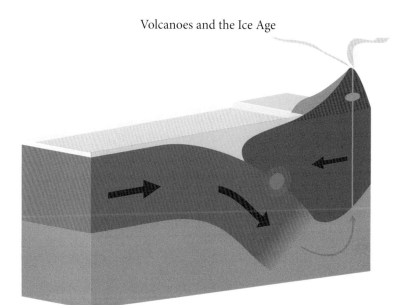

Plate tectonics

For example, the 1815 eruption of Mount Tambora in Indonesia caused the "year without a summer" across Europe in 1816. Michael Oard refers to this temporary cooling of Earth as the "anti-greenhouse" effect.[4] However, that particular cooling resulted from a single large explosive eruption. An Ice Age needs sustained eruptions of that kind over many decades or even centuries.

Not just any volcanism would accomplish this. Volcanoes are not all the same. The most common types of volcanoes across most of the ocean basins have basalt-rich magmas and are less capable of producing the explosions necessary to generate sun-blocking ash and aerosols (tiny particles or droplets).[5] That kind of explosion needs the specific, volatile, silica-rich magmas generated by partial melts at subduction zones. So, what could cause enough of this kind of volcanic activity to produce an Ice Age?

The answer is the rapid subduction involved in catastrophic plate tectonics. Stratovolcanoes form above subduction zones as ocean lithosphere is pulled down into Earth's mantle. The heat of the mantle causes a partial melting of the crust. The first minerals to melt are those with the lowest melting points, such as quartz, feldspar, and biotite—the main components of granite—resulting in a granitic (silica-enriched) magma.

Stratovolcano eruptions often emit large amounts of sulfur dioxide gas. Chemical reactions in the atmosphere form sulfuric acid droplets, which can remain in the stratosphere for two to three years. Subduction also introduces a lot of water into the melt, increasing the volatility of the magma. Granitic melt then rises and erupts catastrophically, sending ash and aerosols high into the atmosphere. The explosive volcanoes that accompanied subduction during the Flood, as well as their continued eruptions through the early post-Flood period, provided the aerosols needed for many years of summer cooling.

The two conditions that result in an Ice Age must be met simultaneously—centuries' worth of silica-rich volcanism to produce aerosols to cool Earth and hot oceans to cause higher evaporation rates and the snowfall necessary to make the massive continental ice sheets. Together, these factors created the perfect conditions for the Ice Age. Then, as the ocean crust and the water above slowly cooled and volcanic activity diminished during the centuries after the Flood, the Ice Age ended.[4]

The subduction process and its results were no coincidence or accident. Conventional science struggles to explain the Ice Age. But for those who affirm the historicity of God's Word, it's no mystery. The Flood described in Genesis provides the framework we need to decipher Earth's past.

References

1. Oard, M. J. 1990. *An Ice Age Caused by the Genesis Flood.* El Cajon, CA: Institute for Creation Research.

2. Clarey, T. 2016. Embracing Catastrophic Plate Tectonics. *Acts & Facts.* 45 (5): 8–11.

3. Hebert, J. 2017. The Flood, Catastrophic Plate Tectonics, and Earth History. *Acts & Facts.* 46 (8): 11–13.

4. Oard, M. J. 2004. *Frozen in Time: The Woolly Mammoth, the Ice Age, and the Bible.* Green Forest, AR: Master Books.

5. Raymond, L. A. 1995. *Petrology: The Study of Igneous, Sedimentary, and Metamorphic Rocks.* Dubuque, IA: William C. Brown Communications.

18
PREDICTING VOLCANIC ERUPTIONS USING MUOGRAPHY

Timothy Clarey, Ph.D.

In 2020, a study published in *Scientific Reports* outlined a novel method to predict volcanic eruptions.[1] However, the technique only seems to work on a site-by-site basis and requires a tremendous amount of eruption data, more information than most volcanoes usually provide.

Volcanic activity is usually monitored by using surface elevation changes, gas seeps, and seismic data. None of these methods can predict the exact day or time of an eruption. The best predictive techniques still leave a great amount of uncertainty, sometimes with an error margin of weeks or months at best.

This method involves the collection of muons. Muons are the result of cosmic rays from the sun. Mara Johnson-Groh explained them this way:

> Cosmic rays continually rain down into Earth's atmosphere from outer space. When they run into atmospheric particles, they decay into smaller components, including an elementary particle called a muon. Muons'

relatively high mass allows them to penetrate deeply into materials, even solid rock. By placing specialized detectors that record muons passing through a volcano, scientists can use the particles to create more finely defined maps of the interior of the volcano than possible with previous techniques.[2]

The result is a visual display of the internal plumbing of the volcano. This technique allows scientists to much more accurately map out the detailed movements of the magma under the volcano compared to previous methods.

Yukihiro Nomura from the University of Tokyo Hospital and his colleagues from various institutions across Japan studied Sakurajima volcano, one of the most active volcanoes in the world between 2014 and 2016, while it experienced multiple eruptions.[1]

Nomura and his co-authors described the technique in their paper:

Muography is a newly developed imaging technique utilizing high-energy near-horizontally arriving cosmic muons and enables us to visualize the internal structures of large objects. Muography produces a projection image (hereafter, muogram) of a large body by mapping out the number of muons that are transmitted through it. Muograms were first used in 1970 by Alvarez *et al.* to search for hidden chambers in the Chephren's Second Pyramid.[1]

The study authors warn that this technique can

The Great Sphinx of Giza and the Pyramid of Chephren

only make accurate predictions for volcanoes that have sufficient historical eruption data. Calibration requires the collection of muograms for numerous eruptive events.[1] Muography worked to predict eruptions of Sakurajima volcano because it had so many eruptions in a timeframe of only a few years. They also noted that every volcano will have to be studied individually over multiple eruptive events, because each magma chamber likely behaves differently. Further studies will also need to compare the muography results with other techniques, like gas emission studies and seismic imaging, to see if these other methods can supplement the accuracy of the muography method.

Rebecca Savage, a volcanologist at the University of Calgary who was not part of the study team, explained:

Forecasting of a volcanic eruption rarely relies upon a single parameter, and therefore, the combined use of monitoring tools and forecasting methods is likely to give the "best" outcome. Since Sakurajima has been well monitored for a long time, it would be interesting to see how muography compares to other, more traditional, monitoring techniques, such as seismicity, deformation, and gas emission, in terms of its ability to successfully forecast an eruption.[2]

Volcanoes were an important part of the end of the Flood year and in the centuries thereafter. The tiny aerosols and ash particles blasted out of subduction zone volcanoes were critical in lowering global temperatures sufficiently to bring on the Ice Age.[3] And without the Ice Age to lower global sea levels by hundreds of feet, there would have been no land bridges to allow animals and humans to migrate from the Ark landing site to the separated continents.[4] God had a plan. Volcanoes today, like Sakurajima volcano, are reminders of a network of active volcanoes that were used by God to bring about His plan.

References

1. Nomura, Y. et al. 2020. Pilot study of eruption forecasting with muography using convolutional neural network. *Science Reports.* 10 (5272).

2. Johnson-Groh, M. Are cosmic rays a key to forecasting volcanic eruptions? *Eos News.* Posted on eos.org April 21, 2020, accessed April 28, 2020.

3. Clarey, T. 2019. Subduction Was Essential for the Ice Age. *Acts & Facts.* 48 (3): 9.

4. Clarey, T. 2020. *Carved in Stone: Geological Evidence of the Worldwide Flood.* Dallas, TX: Institute for Creation Research, 354–375.

19
VOLCANIC ASH TURNS
TO STONE IN MONTHS

Timothy Clarey, Ph.D.

How long does it take for volcanic ash to turn to stone? Most uniformitarian scientists claim this is a slow process that should take many years, even thousands of years. But what does empirical science reveal? Does ash take a long time to harden?

Evidence from the Philippines indicates it takes a lot less time than previously believed. It may take as little as a few weeks or maybe just a few months.

On January 12, 2020, Taal Volcano in the Philippines erupted after 43 years of dormancy. Its ash spread over 60 miles north of the volcano, causing 40,000 people to be evacuated. Over 6,000 people who refused to leave their homes died.[1]

NASA scientists had before and after images of Taal Volcano. They found some dramatic changes in the December 6, 2019, image and the March 11, 2020, image.[1] And when they examined the ash on the ground, they found some other surprising results.

Scientists noted that following the eruption, rain-

Volcanic cone in Taal Lake, Philippines

fall wet the freshly fallen ash and transformed it into a mud-like substance that soon hardened into something like cement.[1]

Denison University volcanologist Erik Klemetti said, "Most of the ash that fell within the caldera is in the process of getting concentrated into gullies and streams or deposited into the lake."[1] Stacy Liberatore at the *Daily Mail* reported,

> NASA explains ash not blown or washed away became wet following the months after the eruption, which turned into a mud-like texture and hardened into something similar to cement—trapping it on the island.[1]

Scientists have already discovered rapidly-forming natural cement in other places. A company called CarbFix has been working with the Hellisheidi Power station near Reykjavik, Iceland, to conduct carbon capture experiments.[2] They found that sedimentary rock, like limestone, can also form quickly.[2] CarbFix plant manager Dr. Edda Sif Aradottir explains,

> The process starts with the capture of waste CO2 from the steam [of the hydrothermal plant], which is then dissolved into large volumes of water. We use a giant soda-machine. Essentially what happens here is similar to the process in your kitchen, when you are making yourself some sparkling water: we add fiz to the water.[2]

Then the carbonated water is transferred to a deep well injection site about a mile away and pumped into the local rocks at depths of about 3,200 feet down.[2]

The local rock in Iceland is a volcanic rock called basalt, one of the most common rocks on Earth. Basalt that cools near the surface often contains a high percentage of rounded holes caused by gas bubbles in the cooling magma. The result is a fairly porous rock that resembles Swiss cheese. As the injected CO_2-rich water percolates through the holes in the basalt, it dissolves some of the calcium and magnesium from the basalt and precipitates minerals that make limestone.

A little over a year after starting the CO_2 injection, CarbFix drilled down and cored the surrounding rock near their injection well. What they found surprised them. The basalt rock was speckled with white minerals. Nearly all of the holes were filled with carbonate minerals.

The amazing thing is how fast this process took. Sigurdur Gislason of the University of Iceland explains,

> Before the injection started in CarbFix, the consensus within the [conventional] scientific community was that it would take decades to thousands of years for the injected CO_2 to mineralise. Then we found out that it was already mineralised after 400 days.[2]

Uniformitarian beliefs fail again. Rocks don't need thousands or millions of years to form. The process simply requires the right conditions. These two studies show that rock does form quickly, easily within the timeframe of the global Flood described in Genesis. Thousands or millions of years are not necessary. The present study reminds us that conventional notions of deep time are not based on observation but assumption.[1]

References

1. Liberatore, S. Incredible NASA images reveals ash damage from the Philippines' Taal Volcano eruption that transformed the tropical terrain into 'the moon's surface.' *Daily Mail.* Posted on dailymail.co.uk March 18, 2020, accessed March 25, 2020.
2. Perasso, V. Turning carbon dioxide into rock—forever. *BBC News.* Posted on bbc.com May 18, 2018, accessed March 25, 2020.

Agung eruption

20
MASSIVE RELEASES OF CO_2 FROM VOLCANISM RIVAL HUMANS

Timothy Clarey, Ph.D.

A 2020 study published in *Nature Communications* suggested that pulses of massive amounts of lava can release as much CO_2 as humanity will produce for the entire 21st century. This indicates that volcanic activity, especially during the global Flood and right after, likely produced tremendous amounts of CO_2 that has far outweighed any produced by humans.

This really should be no surprise, because today's volcanoes still produce vast amounts of CO_2 and water. However, these scientists were able to find evidence of vast quantities of ancient CO_2 still trapped in the rocks themselves.

Manfredo Capriolo from the University of Padova, Italy, and a consortium of scientists from institutions across Europe and Morocco collected over 200 samples of igneous rock (formerly molten) from North America, Africa, and Western Europe. These rocks were all part of the similarly-aged Central Atlantic Magmatic Province (CAMP) that intruded and produced extensive lavas across eastern North America, South America, western Africa, and even part of Southeast Europe.

This CAMP volcanic activity is associated with the initial break-up of the supercontinent of Pangaea and the beginning of the modern Atlantic Ocean. Conventional scientists claim the rocks are about 201 million years old, falling right at the end of the Triassic in the rock record.[1]

Furthermore, uniformitarian scientists assert that the end of the Triassic coincided with a massive extinction event. They claim that the end of the Triassic is one of the "big five" extinctions in geologic history. Other extinctions include the Permian-Triassic and the Cretaceous-Paleogene, which are both thought to have been even bigger extinctions.

Capriolo and his co-authors found carbon dioxide and water vapor-bearing bubbles in about 20 of the 200 samples.[1] Few other volatiles (vapor-rich) compounds were identified. They discovered that the CAMP igneous rocks were all rich in CO_2.[1] The authors wrote:

The CO_2-bearing bubbles identified in CAMP MIs [melt inclusions in the minerals] can be interpreted as batches of ascending volatiles [bubbles] entrapped in crystalline mush shortly prior to its mobilization and prior to its eruption....For instance, CO_2-rich Hawaiian basalts have been shown to rapidly rise from over 5 km [3.1 miles] depth and to cause high fountaining eruptions.[1]

Capriolo and his colleagues totaled up the volume of CO_2 released and concluded,

> Our estimates suggest that the amount of CO_2 that each CAMP magmatic pulse injected into the end-Triassic atmosphere is comparable to the amount of anthropogenic emissions projected for the 21st century.[1]

They went on to speculate that this massive release of CO_2 during the formation of the CAMP could be a main cause of the end-Triassic extinction. Hannah Osborne reported,

> David Bond, from the U.K.'s University of Hull, whose research focuses on environmental changes during mass extinctions and who was not involved in the latest study, commented on the findings....

> "There is little doubt that some LIPs [Large Igneous Provinces] cause mass extinctions. However, there are examples of huge LIPs that do not seem to have given a 'deadly kiss' to life on Earth. This paper by Capriolo and colleagues helps explain why some LIPs are deadly, while others are not. They look inside the anatomy of a killer, through a case study of the enormous CAMP."[2]

But are these real extinctions? Is CO_2 the "killer" it is being made out to be? And why don't all large igneous provinces cause extinctions?

In a young-earth Flood model, there are no extinctions in the rock record, only last occurrences.[3]

All fossils suddenly appear at some level, continue for a while unchanged, and disappear suddenly at a higher level (the claimed extinction).

Creation geologists believe much of the rock record was deposited rapidly in the year-long Genesis Flood. The end-Triassic is just a sudden change in fossil types as the floodwaters inundated higher and new levels of the pre-Flood world.[4]

A pre-Flood world with different ecological zones that were simultaneously flooded globally explains what we observe in the fossil record.[4] It is likely that very few land animals went extinct until after the Flood.

Uniformitarian scientists struggle to make sense of the fossil record and the causes for what they believe are extinctions. The end-Triassic "extinction" is no different. And CO_2 is produced by most volcanoes. It is only the uniformitarian worldview that keeps them searching for a cause.

The answer is in plain sight. They just need to open their Bibles to the account of the Flood in the book of Genesis. Then the rocks, fossils, and volcanic activity all make sense. Massive releases of CO_2 are nothing to be feared.

References

1. Capriolo, M. et al. 2020. Deep CO_2 in the end-Triassic Central Atlantic Magmatic Province. *Nature Communications.* 11: 1670.

2. Osborne, H. Projected CO_2 emissions similar to those released by volcanoes during mass extinction even 200 million years ago. *Newsweek.* Posted on newsweek.com April 7, 2020, accessed April 14, 2020.

3. Clarey, T. 2020. *Carved in Stone: Geological Evidence of the Worldwide Flood.* Dallas, TX: Institute for Creation Research, 90–113.

4. Ibid, 400–417.

21
EXPRESS-LANE MAGMA INDICATES YOUNG EARTH

Timothy Clarey, Ph.D.

Magma can really make tracks according to a 2013 study published in *Nature* that significantly upped the perceived speed limit of magma movement in the earth.[1] Philipp Ruprecht and Terry Plank examined rocks generated from the most recent, 1963–1965, eruption of the Irazú volcano in Costa Rica. They found indications that the magma traveled at 150–300 feet per day and possibly as fast as several thousand feet per day. These findings present problems for old-earth theories that are based on sluggish magma movement.

Olivine crystals found in the rocks generated during this latest eruption were key factors in the study. These crystals preserved the chemical signature of the mantle below the crust from as deep as 22 miles below the surface. Lack of chemical mixing within the magma below the volcano indicates that ascent times were extremely short, with travel from the top of the mantle to the surface taking only a matter of months.

Magmas ascend through the earth because the liquid is more buoyant than the surrounding rocks, somewhat like a hot air balloon traveling upward through the cooler air. However, for many decades, most uniformitarian scientists advocated for slow-moving, slow-cooling magmas that inched their way to the surface over thousands or even millions of years.[2] The present study smashes this ingrained myth, replacing it with evidence of a more rapid, "catastrophic" magma ascent.

Creation scientists advocate for this theory of rapid ascent and cooling of magmas.[2,3] Granites in the Front Range of Colorado and in the mountains of British Columbia, Canada, had telling ascent rates

Irazú Volcano, Costa Rica

between one half and nine miles per year.[2] However, these magmas originated in the crust at depths only 13 miles down. The *Nature* study reveals a rapid ascent from depths as far down as the top of the mantle—well beyond 20 miles deep![1]

Conventional scientists argue that the Sierra Nevada batholith in California, a large magma chamber many miles across, formed by slow magma movements over a 40-million-year time span. However, based on the newer ascension-rate data, even extensive granitic batholiths like the Sierra Nevada could have formed in just over 1,000 years.[2]

Studies showing the brisk rise of magma during volcanic eruptions are now becoming more common, and scientists are even considering such movement "catastrophic."[4] *Nature* authors Ruprecht and Plank conclude, "This is not an isolated occurrence; magma mixing, mafic magma recharge, and high Fo [fosterite] olivines are common to many stratovolcanoes above subduction zones [i.e., the Cascade Range in Washington and Oregon], and the approach we have outlined here may be applied generally."[1]

Evidence supporting rapid and catastrophic movement of magma fits the young earth model, proving that millions and billions of years are not necessary to form the geologic features we see today. The reality is that volcanoes and magmas can form and move rapidly, a fact that supports the young age of the earth spelled out in the book of Genesis.

References

1. Ruprecht, P. and T. Plank. 2013. Feeding andesitic eruptions with a high-speed connection from the mantle. *Nature*. 500 (7460): 68–72.

2. Snelling, A. 2009. *Earth's Catastrophic Past, Volume 2.* Dallas, TX: Institute for Creation Research, 987–993.

3. Woodmorappe, J. 2001. The rapid formation of granitic rocks: more evidence. *Journal of Creation.* 15 (2): 122–125.

4. Petford, N. et al. 2000. Granite magma formation, transport and emplacement in the earth's crust. *Nature.* 408 (6813): 669–673.

22
HALEAKALA NATIONAL PARK: ONE OF MANY YOUNG-LOOKING VOLCANOES

Brian Thomas, Ph.D.

You can start the day atop the cold peak of Mt. Haleakala and end it on a warm beach on Maui. The sprawling volcanic mountain rises 10,000 feet above sea level. Nēnē birds, cattle, horses, and people populate its slopes and summits. When did Haleakala form?

Maui formed after Oahu and before the Big Island. Hawaiian schoolchildren memorize the standard age assignments. From west to east, Kauai formed supposedly about five million years ago, Oahu three million, Maui Nui one million, and the Big Island 0.4 million. The Big Island is still forming today.[1] But three unique

The Nēnē (Branta sandvicensis), a variety of goose, is the state bird of Hawaii

Caldera of the Haleakala volcano, Hawaii

observations in and around Haleakala National Park make the volcano look only thousands of years old, not millions.

Circular Reasoning with Hawaiian Stick Spiders

Spider studies suggest the first clue to a more recently formed Maui. Elusive nocturnal Hawaiian stick spiders creep among the tropical forests on the mountain's north side. Biologists marvel at how fast these arthropods switch between different body colors. Gold ones live on leaves, white ones on lichen, and dark ones among rocks.

Patterns in their DNA suggest that they pioneered the Hawaiian Islands moving east, starting at Kauai (Figure 1). When a dark stick spider lands on a new island, its offspring suddenly display dark, gold, or white forms. They deploy their color changes fast-

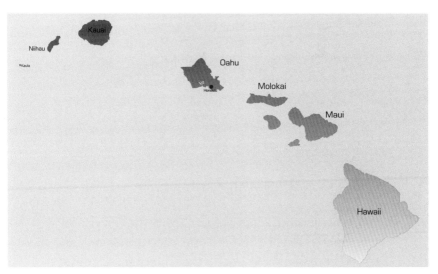

Figure 1. *Hawaiian Island chain*

Author searches for Hawaiian stick spiders

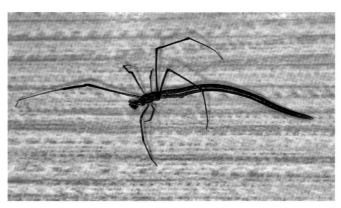

Hawaiian stick spider (genus Ariamnes)

er than "gold and white spiders from the old island have time to reach the new one," according to *Cell Press.*[2] These fast flips suggest that internal programming, not many generations of death, sits at the helm

118

of these spider adaptations.

Authors publishing in the journal *Current Biology* framed a spider history using the standard times assigned to the islands as an initial assumption.[3] This sounds like circular reasoning.

They offered a defense, writing, "Because of the inevitable circularity of using island age for calibration, we applied a general arthropod molecular clock estimate of 2.3%/myr."[3] Then they referenced a 1994 study as a basis for their "clock." Problem solved? Not at all. The author of that 1994 study wrote, "The estimates of divergence times for these [arthropod] taxa [kinds] are based on dated geological events reported by the authors." Of course, the use of "dated" geological events is just as circular as using "island age" to calibrate timelines, since "island age" was determined by "dated" geological events.

Circularity is a poor substitute for good science but opens the door to considering a second observation that casts even more doubt on standard dating.

Wrong Isotope Ages for Volcanic Rocks

Conventional scientists use radioactive isotopes to estimate ages for certain rocks. The process requires workers to assume how much of which isotope was present when the rock cooled. Some classic tests of this assumption used Hawaiian lava rocks. The tests were simple: compare radioisotope age estimates to actual recorded ages. Table 1 shows some results.[4]

Just like these Hawaiian lava rocks, isotope age estimates fail to capture known ages for rocks found

Sample	Location	Isotope Age	Actual Age	Reference
Hualalai basalt	Hawaii	1.6–0.19 Ma*	AD 1800–1801	5
Hualalai basalt	Hawaii	22.8–16.5 Ma	AD 1800–1801	6
Kilauea Ika basalt	Hawaii	8.5–6.8 Ma	AD 1959	6
Kilauea basalt	Hawaii	21–8 Ma	< 200 years old	7

Table 1

* "Ma" is millions of years.

around the world. Clearly, we need a whole new way to date rocks. What if we built our world history on what the Lord Jesus—who has always been there—says in His Word? The Bible-based alternative of a flood that reshaped the globe may sound silly to those who first hear it, but it fits our next observation: landforms.

Author faces Haleakala while investigating lava rocks

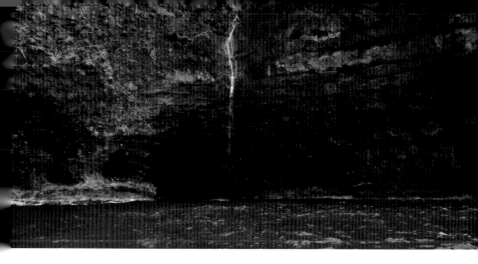

Flat-lying lava layers, uncollapsed lava tubes, and steep cliffsides on Kauai—the oldest of the Hawaiian Islands—signal recent and rapid formation for all the islands, including Maui

Three Youthful Landforms

A hike along Haleakala's Halemau'u Trail reveals flat-lying lava layers. All the islands have these. It looks like the lava landed in sheets, one right after the other. If eons elapsed on Maui between these layers, then where are the ruts and valleys that normal erosion would have etched into the top of each lava layer?

Visitors enjoy Hana Lava Tube northeast of Haleakala. All the islands have lava tubes. The brittle, hollow structures should collapse after only hundreds of thousands of years. The islands and their tubes must have arisen when new seafloors formed near the end of Noah's Flood only thousands of years ago.[8]

We peered over steep cliffsides near the top of Haleakala. Steep-walled cliffs erode faster than other landforms. Indeed, the measured erosion rate of five inches per year would have rounded cliffs and flattened whole islands after a million years.[9] How could these islands keep their cliffs so long?

If you visit Haleakala National Park, remember how circular reasoning in spider ages and isotope dating failures both cast doubt on millions-of-years thinking. Flat-lying lava layers, fragile lava tubes, and steep cliffs with measured erosion rates point to a Bible-friendly age of thousands of years for the whole world, including its volcanoes.

References

1. Jones, C. Hawaii's Kilauea volcano erupts, lava fountains form in park. Associated Press. Posted on APnews.com September 29, 2021, accessed October 1, 2021.

2. Hawaiian stick spiders re-evolve the same three guises every time they island hop. *Cell Press.* Posted on ScienceDaily.com March 8, 2018, accessed October 1, 2021.

3. Gillespie, R. G. et al. 2018. Repeated Diversification of Ecomorphs in Hawaiian Stick Spiders. *Current Biology.* 28: 941–947, referencing Brower, A. V. 1994. Rapid morphological radiation and convergence among races of the butterfly *Heliconius erato* inferred from patterns of mitochondrial DNA evolution. *Proceedings of the National Academy of Sciences.* 91 (14): 6491–6495.

4. Snelling, A. A. 1998. Andesite Flows at Mt. Ngauruhoe, New Zealand, and the Implications for Potassium-argon "Dating." *Proceedings of the Fourth International Conference on Creationism.* Pittsburgh, PA: Creation Science Fellowship.

5. Dalrymple, G. B. 1969. ^{40}Ar/^{36}Ar Analyses of Historic Lava Flows. *Earth and Planetary Science Letters.* 6: 47–55.

6. Krummenacher, D. 1970. Isotopic Composition of Argon in Modern Surface Volcanic Rocks. *Earth and Planetary Science Letters.* 8: 109–117.

7. Noble, C. S. and J. J. Naughton, 1968. Deep-Ocean Basalts: Inert Gas Content and Uncertainties in Age Dating. *Science.* 162: 265–267.

8. Austin, S. A. et al. 1994. Catastrophic Plate Tectonics: A Global Flood Model of Earth History. In *Proceedings of the Third International Conference on Creationism.* R. E. Walsh, ed. Pittsburgh, PA: Creation Science Fellowship.

9. Fletcher, C. H. et al. National Assessment of Shoreline Change: Historical Shoreline Change in the Hawaiian Islands. U.S. Geological Survey Open-File Report, 2011–1051, 55.

23
MINISCULE EROSION POINTS TO HAWAII'S YOUTH

Timothy Clarey, Ph.D.

Conventional scientists claim the Hawaiian Islands are millions of years old based primarily on radioisotope dating. Yet, the landforms and measured erosion rates tell a far different story—a story that better matches the Bible.

The Hawaiian Islands are a chain of islands in the middle of the Pacific Ocean on the Pacific plate (Figure 1). The conventional explanation for them is

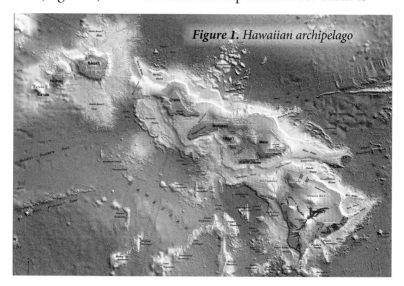

Figure 1. Hawaiian archipelago

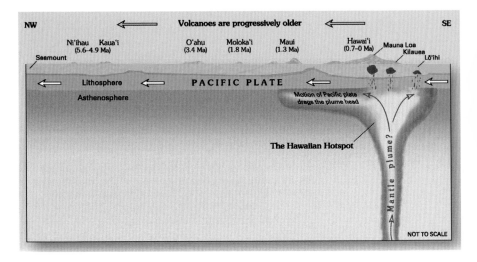

Figure 2. *Hawaii hotspot*

that they formed as a result of volcanic activity when the plate passed over a hotspot in the mantle at a rate of inches per year (Figure 2). As the islands moved off the hotspot and their volcanoes became inactive, they left a trail of progressively older volcanic islands in the northwesterly direction of plate motion (Figure 1).

In contrast, creation geologists attribute the formation of the islands to recent volcanic activity during the Flood. Geophysicist John Baumgardner demonstrated that the plates would have moved much more quickly in the Flood—at rates of several yards per second—creating the Hawaiian Islands just thousands of years ago.[1]

The rocks and landforms of Hawaii also tell a different story from the conventional version. Lava tubes and waterfalls, common on all the islands, are evidence of youth. Lava tubes form as natural conduits to transport molten lava, but today they are merely hollow, cave-like "pipes." These tubes cannot exist for millions of years without collapsing. Steep valleys, steps, and waterfalls should have eroded away long ago, forming a gentle, subdued landscape over the course of millions of years. Yet, we still see lava tubes, steep valleys, and dramatic waterfalls on all the islands.

We also observe extensive layers of lava on every island. Stacked lava layers are evidence of rapid volcanic deposition, placing layer upon layer with no evidence of time or erosion between any flow.

But the strongest evidence for a youthful Hawaii comes from the measured erosion rates along the coastlines of the islands.[2] Scientists studying photographs and maps since 1900 found that most beaches on Kauai, Oahu, and Maui experience erosion averaging 0.4 feet/year, or about five inches per year. United States Geological Survey Director Marcia McNutt explains:

> The inevitable fate of the Hawaiian Islands millions of years into the future is seen to the northwest in the spires of French Frigate Shoals and the remnants of other once mighty islands, ancestors of today's Hawaii, but now sunken beneath the sea through the forces of waves, rivers, and the slow subsidence of the seafloor.[3]

This erosion process would completely destroy the islands in only a few hundred thousand years. Doing the math, we get 76 miles of erosion in only one million years (at 0.4 ft/yr), which would completely eliminate the islands—except possibly the Big Island where volcanism is still occurring since it currently sits on the hotspot. If the islands are really millions of years old, they should have eroded beneath the sea long ago.

Conventional science cannot claim there was not an ocean around the islands to prevent erosion in the past. Nor can they claim renewed uplift and mountain building as a rescuing device to preserve the islands.[4] As each of the islands move off the hotspot (Figure 2), they cool, sink, and rapidly erode away in just thousands of years. Once off the hotspot, there is simply no new lava source to keep them "afloat."

However, if they are only around 4,500 years old, then the islands have experienced about a third of a mile of erosion. And that is precisely what we observe. The Hawaiian Islands really are young.

References

1. Baumgardner, J. 2003. Catastrophic Plate Tectonics: The Physics behind the Genesis Flood. In *Proceedings of the Fifth International Conference on Creationism*. R. L. Ivey Jr., ed., Pittsburgh, PA: Creation Science Fellowship, Inc., 113–126.

2. Fletcher, C. H. et al. National Assessment of Shoreline Change: Historical Shoreline Change in the Hawaiian Islands. U.S. Geological Survey Open-File Report, 2011–1051, 55.

3. 70 Percent of Beaches Eroding on Hawaiian Islands Kauai, Oahu, and Maui. *USGS Newsroom*. Posted on usgs.gov May 7, 2012, accessed November 1, 2016.

4. Thomas, B. Continents Should Have Eroded Long Ago. *Creation Science Update*. Posted on ICR.org August 22, 2011, accessed November 1, 2016.

24

CRATER LAKE NATIONAL PARK: SERENE BEAUTY AFTER VOLCANIC HISTORY

Timothy Clarey, Ph.D.,
and James J. S. Johnson, J.D., Th.D.

The deepest and arguably most spectacular lake in the United States is inside a volcano. Known as Crater Lake, it reaches 1,943 feet at its deepest point.[1] Crater Lake National Park was established in 1902 by President Theodore Roosevelt and became America's fifth national park. Near the center of the Cascade Mountains in south-central Oregon, Crater Lake is about a five-hour drive south from Portland and is eight hours north of San Francisco.[2]

How and when did this lake form? The evolutionary story talks about the Cascade Mountains beginning many millions of years ago.[1] However, this date is based on disproven or unverifiable assumptions and evolutionary dogma. The real account begins with the global Flood just 4,500 years ago.

Flood Origin of Crater Lake

In ICR's Flood model, the Cascade Mountains began development during the receding phase of the Flood.[3] The source magmas were generated as the Pacific seafloor was rapidly subducted or pulled under the West Coast of North America during the Flood year.[4] This process created explosive magmas much different from the Hawaiian volcanoes[5] and caused repeated eruptions late in the Flood year and into the Ice Age. The Cascades grew very quickly into massive volcanoes.

Yellow-bellied marmot

The volcano beneath Crater Lake catastrophically erupted for a final time during the Ice Age. Formerly known as Mount Mazama, the 13,000-foot-high vol-

cano blasted out about 75 cubic miles of material.[2] This caused it to collapse upon itself, creating a bowl-shaped crater (caldera). This eruption was 42 times more powerful than the 1980 eruption of Mount St. Helens.[2] The highest elevation within the park is now just 8,928 feet at Mount Scott.[1]

Following this final catastrophic explosion, a few lavas flowed out into the open caldera, and finally a tall cinder cone developed. Known as Wizard Island for its cone-shaped appearance, the island rises almost 800 feet above the lake's surface.[1] Cinder cones often form during a final degassing episode (like steam) as volcanoes go dormant or extinct.[6]

Early visitors in 1853 called it Deep Blue Lake, but it was later called Crater Lake in a newspaper account.[1] No streams flow into the lake, and it is only fed by snowmelt from winter snowfalls of about 45

Crater Lake

feet each year.[2] Water is only removed by evaporation and groundwater seepage.[1] Because of these factors, the lake is one of the clearest in the world, with visibility down to 120 feet.[1]

Exhibiting Christ's Glory Today

Crater Lake exhibits year-round beauty for those with eyes to see it, and about a half-million visitors view it each year. But some of Christ's creatures don't just visit, they call it home. As a clean freshwater lake, Crater Lake's pure water is a habitat for landlocked kokanee salmon and rainbow trout.

Its surrounding shorelands and Wizard Island provide an evergreen-forested and felsenmeer habitat for a variety of large animals—"black bear, bobcat, deer, and marmots"[2]—plus many smaller ani-

Merganser duck

mals such as pika, chipmunks, and golden-mantled ground squirrels.[7]

Birdwatchers, too, enjoy visiting Crater Lake. Depending upon the time of year, Crater Lake hosts many migratory or resident birds.

> Bald eagles (*Haliaeetus leucocephalos*) and peregrine falcons (*Falco peregrinus*) nest along the caldera cliffs. American dippers (*Cinclus mexicanus*, America's only aquatic songbird) forage at the bottom of fast-flowing streams. Subalpine areas are home to the gray-crowned rosy finch (*Leucosticte tephrocotis*). Wildfire burned forests attract a variety of woodpeckers….Common mergansers (*Mergus merganser*) raise families on the lake, and calls of songbirds permeate the forests and meadows.[7]

Who would expect such a beautiful lake surrounding such a magnificent volcanic cinder cone island? Those who visit Crater Lake and who gaze on Wizard Island can echo the prophet Isaiah's words: "Let them give glory to the Lord, and declare His praise in the coastlands" (Isaiah 42:12).

References

1. Hopson, R. F. 2018. Crater Lake National Park, Southwest Oregon. In *The Geology of National Parks*, 7th ed. D. Hacker, D. Foster, and A. G. Harris, eds. Dubuque, IA: Kendall-Hunt, 649–666.

2. Macy, M. 1999. Crater Lake. In *America's Spectacular National Parks*. L. B. O'Connor and D. Levy, eds. Los Angeles, CA: Perpetua Press, 120–121.

3. Clarey, T. 2020. *Carved in Stone: Geological Evidence of the Global Flood*. Dallas, TX: Institute for Creation Research.

4. Clarey, T. Plate Subduction Beneath China Verifies Rapid Subduction. *Creation Science Update*. Posted on ICR.org December 23, 2020, accessed October 29, 2021.

5. Clarey, T. 2019. Subduction Was Essential for the Ice Age. *Acts & Facts.* 48 (3): 9.

6. Cinder cones are made of a rock called scoria. These are volcanic rocks that contain many holes from gas bubbles. Cinders are about the same size as lava rocks for landscaping and/or gas grills, around 0.08–2.5 inches in length.

7. Birds. Crater Lake National Park, Oregon. National Park Service. Posted on nps.gov, accessed November 1, 2021. Regarding animals in parks and wildlife refuges of the Great West, see also Johnson, J. J. S. Yes, Deer, It's Time for Some Calm News. *Creation Science Update.* Posted on ICR.org April 27, 2020, accessed October 29, 2021. This chapter's coauthor, Dr. Johnson, visited Crater Lake and Lava Butte, Oregon, during the summer of 1992 for biogeography research for use in teaching ecology and ornithology at Dallas Christian College.

25
DISCOVERY: VOLCANOES ON VENUS

Brian Thomas, Ph.D.

The tortured surface of Venus appears to have been formed through recent geologic processes, and its rocks contain no record of deep time.[1] What if Venus were young rather than four and a half billion years old? It would explain quite a bit, including a discovery made by scientists peering through its dense atmosphere.

Gathering clues from Venus' cloud-covered surface is no easy task. Astronomers based at Brown University stitched together 2,463 images of a rift system called Gani-ki Chasma taken by the Venus Express spacecraft as it orbited the planet. The astronomers created a time-lapse mosaic of

Venus. The volcano Maat Mons, located within the black square, shows signs of a recent eruption.

the rift system and saw intriguing spots that would suddenly burn bright and then quickly fade.

In their technical report, published in *Geophysical Research Letters*, scientists explained that those illuminated spots stayed in one place on the rift system. This showed they did not originate in the atmosphere, since Venus' horrendous weather patterns feature fast-blowing winds of toxic gases that would have dragged any airborne hotspots across the viewing field.[2] And because the stationary bright spots were positioned at the edges of rifts—exactly where volcanic activity on Earth occurs—the researchers concluded they were indeed looking at volcanic activity occurring on Venus.

In a Brown University press release, senior author James Head said,

> We knew that Ganiki Chasma was the result of volcanism that had occurred fairly recently in geological terms, but we didn't know if it

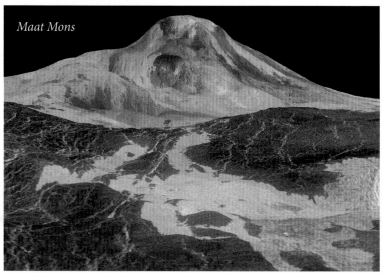

Maat Mons

formed yesterday or was a billion years old. The active anomalies detected by Venus Express fall exactly where we had mapped these relatively young deposits and suggest ongoing activity.[3]

Their data showed that the volcanic eruptions of lava, "become brighter on the time scale of days."[3]

Volcanoes on Venus fit the theme of youthful geologic processes found all around the solar system. They include the excess internal heat of Jupiter, ice-plume geysers from Saturn's moon Enceladus, jet emissions from comets, vaporizing comets, fading planetary magnetic fields, and active volcanism, such as on Jupiter's moon Io and now apparently on Venus.

All of these are difficult to explain if the solar system's planets and comets formed billions of years ago. They should be old, cold, and dead, but instead they're vigorous. What could possibly keep these objects hot, fueled, magnetic, or volcanic after all this time?

Removing assumptions of millions of years solves the fascinating mystery of volcanoes on Venus. The tremendous energy remains in the planet from its creation only thousands of years ago.

References

1. Thomas, B. Most of Venus' History Is Missing? *Creation Science Update.* Posted on ICR.org January 31, 2014, accessed June 24, 2015.

2. Shalygin, E. V. et al. Active volcanism on Venus in the Ganiki Chasma rift zone. *Geophysical Research Letters.* Posted online before print, May 23, 2015, accessed June 24, 2015.

3. Stacey, K. Study suggests active volcanism on Venus. *News from Brown.* Posted on news.brown.edu June 18, 2015, accessed June 22, 2018.

Jupiter's moon Io is the most volcanically active world in our solar system.

26
DESPITE "MAGMA OCEAN" DISCOVERY, IO'S VOLCANIC HEAT REMAINS A MYSTERY

Brian Thomas, Ph.D.

Jupiter's moon Io may have a very short name, but it definitely has the highest volcanic activity of any object in the solar system. Littered with volcanoes, its unique surface sports a massive active volcano named Loki, whose lava output exceeds that of all of Earth's volcanoes put together. Of course, where there's smoke, there's fire, and the origin of Io's volcanic heat remains a mystery.

A creation astronomy video summarized the issue by stating, "If Io is young, it could still be cooling off from its initial formation. But if it's really billions of years old, that energy would have dissipated long ago."[1] Measurements confirmed what researchers had expected for quite a while: Io has a molten or partly molten magma "ocean" beneath its crust.[2]

The study, published in the journal *Science*, found that magnetic signatures detected by the Galileo spacecraft cannot be explained if Io is entirely solid. But a 30-mile-thick molten or partly molten layer lying beneath the moon's crust would account for the magnetic data.[3] Although the question of Io's volcanic lava source has been answered, the source of Io's heat has not been found.

In 1982, authors Pearl and Sinton wrote in *Satellites of Jupiter*, a compilation of research on Jupiter's moons, "Complete elucidation of the heat source remains a significant outstanding problem resulting from the discovery of active volcanism on Io."[4] And since then, no long-age models have been able to match the measured heat output of this tiny, fiery sphere.

Specifically, Io emits 10^{14} watts of power, approximately equal to the output of 10 trillion light bulbs. Virtually all attempts to model Io's heat refer to its "tidal friction." This occurs when Io is pulled in different directions by the gravitational attraction of Jupiter on one side and the planet's other moons on the other side. This causes flexing that generates heat beneath Io's crust, although not enough to account for its massive heat output.

A review paper on the Io heat problem referenced German planet scientist Tilman Spohn, who "acknowledges that there is a gap of about one order of magnitude between the observed heat flow from infrared measurements and the heat flow theoretically determined from tidal [friction] dissipation models."[5] Io's heat output is therefore around *10 times* greater than the long-age models say it should be.

The lead author of the *Science* study said in a Jet Propulsion Laboratory news release, "Scientists are excited we finally understand where Io's magma is coming from and have an explanation for some of the mysterious signatures we saw in some of the Galileo's magnetic field data."[2] But this insight about magma only serves to confirm that Io's interior is hot enough to melt rock! Io continues to torch "billions of years" theories that end in shoulder-shrugging non-explanations for the planet's heat problem. Like so many other objects in the solar system, Io looks quite young.

References

1. *What You Aren't Being Told about Astronomy, Volume 1: Our Created Solar System*. 2009. DVD. Directed by Spike Psarris. Creation Astronomy Media.

2. Galileo Data Reveal Magma Ocean Under Jupiter Moon. Jet Propulsion Laboratory News & Features. Posted on jpl.nasa.gov May 12, 2011, accessed May 17, 2011.

3. Khurana, K. K. et al. Evidence of a Global Magma Ocean in Io's Interior. *Science Express*. Posted on sciencemag.org May 12, 2011, accessed May 17, 2011.

4. Pearl, J. C., and W. M. Sinton. 1982. Hot Spots of Io. *Satellites of Jupiter*. D. Morrison, ed. Tucson, AZ: Arizona University Press, 724–755. Cited in Spencer, W. 2003. Tidal Dissipation and the Age of Io. In *Proceedings of the Fifth International Conference on Creationism*. R. L. Ivey, ed. Pittsburgh, PA: Creation Science Fellowship, Inc., 585–595.

5. Spencer, W. 2003. Tidal Dissipation and the Age of Io. In *Proceedings of the Fifth International Conference on Creationism*. R. L. Ivey, ed. Pittsburgh, PA: Creation Science Fellowship, Inc., 585–595.

APPENDIX
CATASTROPHIC PLATE TECTONICS AND THE FLOOD

Jake Hebert, Ph.D., and Timothy Clarey, Ph.D.

Some Christians hesitate to embrace the notion that the earth's outer surface is moving—and moved even more dramatically during the Flood year. However, tremendous amounts of empirical data suggest significant plate movement occurred just thousands of years ago.[1] Much of these data are independent of conventional deep time and the geologic timescale. In addition, the catastrophic plate tectonics (CPT) model offers a mechanism for the flooding of the continents, the subsequent lowering and draining of the floodwaters, and a cause for the post-Flood Ice Age.

The Flood also enables us to make sense of clues contained within Earth's interior. Our planet can be divided into a thin outer crust, a core at its center, and the mantle between them (Figure 1). The core is comprised of a solid inner core and a liquid outer core. The uppermost part of the mantle and the crust together comprise the lithosphere, about 62 miles thick. Like a cracked eggshell, the lithosphere is divided into seven or eight large plates and many smaller plates.

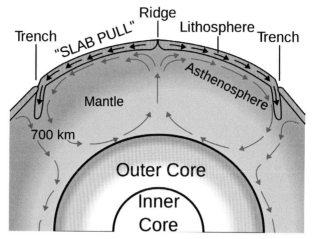

Figure 1. Diagram showing the earth's interor as well as mid-ocean ridge and two subducting slabs

Continental Drift to Plate Tectonics

Geologists derive the theory of plate tectonics from much data collected over many decades. In the early 20th century, Alfred Wegener examined how the continents seem to fit together like a puzzle and matched fossils and mountain ranges across vast oceans to suggest that the continents had split in the past. At the time, his ideas were ridiculed and ignored.

It was not until the 1960s, after immense quantities of oceanographic data were collected, including the publication of Harry Hess' hypothesis of seafloor spreading[2] and J. Tuzo Wilson's early work on plate tectonics,[3] that conventional geologists slowly accepted these ideas. Nearly 50 years after Wegener first proposed the concept of continental drift, the conventional community was overwhelmed with empirical data and reluctantly acknowledged plate tectonics.

Rapid Seafloor Spreading and Runaway Subduction

If continents split, we should find evidence to support these movements under the oceans. In the 1950s and 1960s, geologists discovered that the ocean crust is very young compared to many of the rocks on the continents. In fact, the oldest ocean crust goes back to a brief episode in the Flood during the deposition of the Jurassic system. And at every ridge, the crust gets systematically older in both directions. Although conventional ocean floor maps claim ages of millions of years, they do seem to be correct in a relative sense. Older age dates usually indicate older rocks. In addition, a tremendous amount of data affirms seafloor spreading independent of absolute dating methods.

Creation geophysicist John Baumgardner—described as "the world's pre-eminent expert in the design of computer models for geophysical convection"[4]—has spent many years studying the connection between plate tectonics and the Flood. Today the plates are moving very slowly at rates of just a few centimeters per year, but Dr. Baumgardner argues that they moved much faster in the recent past.[5]

When an oceanic plate and a continental plate collide, the denser rocks of the ocean floor tend to slide under the less-dense continental rocks, a process called subduction. As a subducting plate moves down through the mantle, the resulting friction heats the surrounding material. This heating reduces the viscosity of the material, enabling the subducting plate to move more quickly. As long as the heat is carried away by the surrounding mantle rocks faster than it is generated by the subducting slab, subduction will be slow and gradual.

If, however, the generated heat is not carried away at a sufficient rate, the viscosity of the slab decreases still further, enabling the slab to descend even faster. This results in an effect called runaway subduction in which the subducting slab moves at speeds of meters per second rather than centimeters per year.[5] Fortunately, conditions for runaway subduction are not currently present in the mantle, but there are good reasons to think such conditions occurred in the past.

An imaging process called seismic tomography shows visible lithospheric slabs of oceanic crust going down hundreds of miles beneath ocean trenches and into subduction zones.[6] These are not merely faults, as some have proposed,[7] but 62-mile-thick slabs of brittle, dense rock descending into the mantle. The cooler temperatures exhibited by these subducted slabs of rock create a thermal dilemma for the conventional and old-earth geologists, who must demonstrate how these slabs remained cold for millions of years. Colder, subducted slabs are best explained by runaway subduction just thousands of years ago during the great Flood.[5]

Runaway Subduction: Logical Consequences

If runaway subduction did occur, then certain things logically follow. Since one expects Earth's volume to remain constant during the subduction process, rapid subduction and the destruction of the old seafloor also imply rapid creation of a new seafloor. This would occur at the mid-ocean ridges, where hot magma rises upward (Figure 1).

The lithosphere above the ridge would stretch and thin, allowing the magma to break through the crust. Dr. Baumgardner thinks the mid-ocean ridges, which encircle Earth like seams on a baseball, were the result. As this hot magma came into contact with cold seawater, the result would have been a long, linear geyser that ejected huge amounts of superheated water into the atmosphere. This may have been the source of the intense rains that fell for the first 40 days and 40 nights of the Flood (Genesis 7:12).

CPT Explains the Sources of Water for the Flood and Where It Went

The Bible plainly states that the "fountains of the great deep were broken up, and the windows of heaven were opened" during the initiation of the Flood (Genesis 7:11). In terms of CPT, the breaking up of the fountains of the great deep may be a description of the rifting that took place at the ocean ridges and even within continents.[8] Obviously, the rainfall described as the opening of the "windows of heaven" must have contributed to the Flood.

Additionally, because newly created oceanic lithosphere is hot, less dense, and more buoyant, the CPT model provides another source for water to completely flood the continents. After its formation at the ridges, the freshly formed, lower-density oceanic lithosphere simply pushed up the top of the seafloor from below, displacing ocean water and forcing it on land. Creation geologist Dr. Andrew Snelling calculated that this elevated seafloor could have raised the global sea level by as much as 1.6 km, greatly helping flood the continents.[9]

Rapid movement of the plates during runaway subduction further supplied tsunami-like waves to wash across the land, helping deposit blanket-type sediments across continents. Recent numerical modeling by Dr. Baumgardner has found that repetitive tsunami waves, caused by rapid plate movement, could result in water accumulation more than a kilometer (0.62 miles) deep on the continents, contributing to the flooding.[10] The runaway subduction model also provides a mechanism to lower the continental crust about two miles in the proximity of the subduction zones, causing more extensive flooding of the land and creating room for thousands of feet of sediment.[5]

Subsequent cooling of the newly created ocean lithosphere later in the Flood year (after Day 150) offers an explanation for the lowering of the floodwaters. The 62-mile-thick ocean lithosphere cooled and sank, lowering the bottom of the oceans and drawing the water back off the continents and into the ocean basins.

CPT Explains Rapid Magnetic Reversals

Molten lava, or magma, contains minerals whose magnetic domains tend to align with the direction of Earth's magnetic field. When the rock cools and hardens, this alignment is "locked" into the volcanic rock. The basaltic rocks on either side of the mid-ocean ridges depict a striped pattern consisting of alternating bands of magnetization that reverse direction as one moves away from the ridge. This striped pattern indicates that Earth's magnetic field has flipped dozens of times, with the north and south magnetic poles trading places.

If a new seafloor rapidly formed during the Genesis Flood, then the fact that these magnetic reversals are recorded in oceanic volcanic rocks (most of which were formed during the Flood) implies that the magnetic reversals must also have occurred rapidly. Uniformitarian scientists found strong evidence for rapid magnetic reversals, although such rapid reversals are very hard for them to explain.[11-13]

Creation physicist D. Russell Humphreys proposed a theory that at least qualitatively explains how such rapid reversals could occur.[14] His mechanism requires strong up-and-down motions of fluids within Earth's liquid outer core due to convection. Such convection might be initiated if a cold subducting plate were to come into contact with the outer core at the core-mantle boundary, which Dr. Baumgardner argues is exactly what happened.[15]

CPT Explains Rapid Erosion and Deposition

As the newly formed ocean floor cooled, its density increased and it sank, allowing the floodwaters to drain off the continents. The rapidly receding waters would have eroded away an enormous amount of sediment. In places where the sediments were relatively thin, the water would have eroded all the sedimentary layers, leaving the original basement rocks exposed.

Huge volumes of fast-moving water would have planed some areas flat, resulting in so-called planation surfaces. Since they are not forming today, these surfaces are difficult for conventional geologists to explain.[16] This extensive erosion implies that huge amounts of sediment would have rapidly been dumped into the ocean basins. The Whopper Sand in the Gulf of Mexico—a complete surprise to uniformitarian scientists—is an example of this massive, sheet-like draining of North America.[17]

CPT Explains the Conditions for the Ice Age

Finally, CPT provides a mechanism for the Ice Age that occurred at the end of the Flood. A hot, newly formed ocean crust would have provided tremendous amounts of heat to the ocean waters above. This would have raised the overall temperature of the ocean and caused a greater amount of evaporation, resulting in staggering amounts of precipitation.[18] The increased volcanic activity from the subduction zone volcanoes within the Ring of Fire and elsewhere late in the Flood would have placed huge volumes of ash and aerosols into the atmosphere, cooling the climate most noticeably in the higher latitudes.[18]

The distinctive magmas generated by the partial melt of subducted ocean lithosphere provide the perfect recipe for explosive, ash-rich eruptions. These types of volcanoes (stratovolcanoes) are highest in silica, making them thicker and more explosive.[19] The net result of hotter oceans and tremendous silica-rich volcanic activity brought on from plate motion would be enough to start a widespread Ice Age.

As commonly observed across the bulk of the ocean basins, basalt-rich magmatic volcanoes (shield volcanoes) do not produce the necessary ash-rich explosions to generate sun-blocking aerosols.[19] Only subduction provides these ash-rich magmas. Finally, as the ocean water slowly cooled and volcanic activity diminished over the centuries after Flood, the Ice Age would have ended as abruptly as it began.[18] In contrast, the currently popular conventional Ice Age theory has serious problems.[20]

Conclusion

Accepting the Genesis Flood as literal history enables researchers to make sense of a huge array of data. Although creation scientists are still working to resolve unanswered questions, the creation-Flood model is much more robust and has much more explanatory power than conventional Earth history stories. Skeptics "willingly forget" (2 Peter 3:5) the reality of the Genesis Flood—not because of a lack of evidence but because of an unwillingness to acknowledge God's Lordship.

References

1. Clarey, T. L. 2016. Empirical Data Support Seafloor Spreading and Catastrophic Plate Tectonics. *Journal of Creation.* 30 (1): 76–82.

2. Hess, H. 1962. History of Ocean Basins. In *Petrologic studies: a volume in honor of A. F. Buddington.* A. Engel, H. James, and B. Leonard, eds. Boulder, CO: Geological Society of America, 599–620.

3. Wilson, J. 1968. A Revolution in Earth Science. *Geotimes.* 13 (10): 10–16.

4. Burr, C. 1997. The Geophysics of God. *U.S. News & World Report.* 122 (23): 55–58.

5. Baumgardner, J. R. 1994. Runaway Subduction as the Driving Mechanism for the Genesis Flood. In *Proceedings of the Third International Conference on Creationism.* R. E. Walsh, ed. Pittsburgh, PA: Creation Science Fellowship, 63–75.

6. Schmandt, B. and F.-C. Lin. 2014. P and S wave tomography of the mantle beneath the United States. *Geophysical Research Letters.* 41: 6342–6349.

7. Brown Jr., W. 2008. *In the Beginning: Compelling Evidence for Creation and the Flood,* 9th ed. Phoenix, AZ: Center for Scientific Creation.

8. Reed, J. 2000. *The North American Midcontinent Rift System: An Interpretation Within the Biblical Worldview*. St. Joseph, MO: Creation Research Society Books.

9. Snelling, A. 2014. Geophysical issues: understanding the origin of the continents, their rock layers and mountains. In *Grappling with the Chronology of the Genesis Flood*. S. Boyd and A. Snelling, eds. Green Forest, AR: Master Books, 111–143.

10. Baumgardner, J. 2016. Numerical Modeling of the Large-Scale Erosion, Sediment Transport, and Deposition Processes of the Genesis Flood. *Answers Research Journal*. 9: 1–24.

11. Coe, R. S., M. Prévot, and P. Camps. 1995. New evidence for extraordinarily rapid change of the geomagnetic field during a reversal. *Nature*. 374 (6524): 687–692.

12. Bogue, S. W. and J. M. G. Glen. 2010. Very rapid geomagnetic field change recorded by the partial remagnetization of a lava flow. *Geophysical Research Letters*. 37 (21): L21308.

13. Sagnotti, L. et al. 2014. Extremely rapid directional change during Matuyama-Brunhes geomagnetic polarity reversal. *Geophysical Journal International*. 199 (2): 1110–1124.

14. Humphreys, D. R. 1990. Physical Mechanism for Reversals of the Earth's Geomagnetic Field During the Flood. In *Proceedings of the Second International Conference on Creationism*. R. E. Walsh and C. L. Brooks, eds. Pittsburgh, PA: Creation Science Fellowship, 129–142.

15. Baumgardner, J. R. 2003. Catastrophic Plate Tectonics: The Physics Behind the Genesis Flood. In *Proceedings of the Fifth International Conference on Creationism*. R. L. Ivey, Jr., ed. Pittsburgh, PA: Creation Science Fellowship, 113–126.

16. Oard, M. 2006. It's plain to see: Flat land surfaces are strong evidence for the Genesis Flood. *Creation*. 28 (2): 34–37.

17. Clarey, T. 2015. The Whopper Sand. *Acts & Facts*. 44 (3): 14.

18. Oard, M. 2004. *Frozen in Time*. Green Forest, AR: Master Books.

19. Raymond, L. 1995. *Petrology: The Study of Igneous, Sedimentary, and Metamorphic Rocks*. Dubuque, IA: William C. Brown Communications.

20. Hebert, J. 'Big Science' Celebrates Invalid Milankovitch Paper. *Creation Science Update*. Posted on ICR.org December 26, 2016, accessed May 16, 2017.

CONTRIBUTORS

Steven A. Austin was a research associate at the Institute for Creation Research and earned his Ph.D. in geology from Pennsylvania State University.

Dr. Timothy Clarey is the director of research at the Institute for Creation Research and earned his Ph.D. in geology from Western Michigan University.

Dr. Jake Hebert is a research scientist at the Institute for Creation Research and earned his Ph.D. in physics from the University of Texas at Dallas.

James J. S. Johnson is associate professor of apologetics and chief academic officer at the Institute for Creation Research.

John D. Morris was the president emeritus of the Institute for Creation Research. He earned his Ph.D. in geological engineering from the University of Oklahoma.

Frank Sherwin is a science news writer and a speaker at the Institute for Creation Research. He earned his M.A. in zoology from the University of Northern Colorado and received an Honorary Doctorate of Science from Pensacola Christian College.

Dr. Brian Thomas is a research scientist at the Institute for Creation Research and earned his Ph.D. in paleobiochemistry from the University of Liverpool.

IMAGE CREDITS

THIS BOOK WAS ADAPTED
FROM THE FOLLOWING MATERIALS

Clarey, T. and F. Sherwin. 2020. Mount St. Helens, Living Laboratory for 40 Years. *Acts & Facts.* 49 (5): 10–13.

Thomas, B. 2019. How Mount St. Helens Refutes Evolution. *Acts & Facts.* 48 (6): 14.

Thomas, B. Remembering Mount St. Helens 35 Years Later. *Creation Science Update.* Posted on ICR.org May 26, 2015.

Thomas, B. 2020. Biological Bounceback at Mount St. Helens. *Acts & Facts.* 49 (8): 14.

Morris, J. 2007. Why Does ICR Study the Mount St. Helen's Eruption? *Acts & Facts.* 36 (5).

Austin, S. 2010. Supervolcanoes and the Mount St. Helens Eruption. *Acts & Facts.* 39 (5): 4–5.

Morris, J. 2012. Volcanoes of the Past. *Acts & Facts.* 41 (6): 15.

Clarey, T. and B. Thomas. 2022. Yellowstone National Park, Part 1: A Flood Supervolcano. *Acts & Facts.* 51 (5): 10–13.

Clarey, T. and B. Thomas. 2022. Yellowstone National Park, Part 2: Canyons and Catastrophe. *Acts & Facts.* 51 (7): 10–12.

Clarey, T. Yellowstone Supervolcano Unlikely to Blow. *Creation Science Update.* Posted on ICR.org March 27, 2020.

Clarey, T. World's Largest Volcano Found Hiding Under the Ocean. *Creation Science Update.* Posted on ICR.org May 22, 2020.

Clarey, T. Deep-Sea Volcano Gives Glimpse of Flood Eruptions. *Creation Science Update.* Posted on ICR.org March 10, 2022.

Thomas, B. 2021. Crater of Diamonds State Park and the Origin of Diamonds. *Acts & Facts.* 50 (8): 14–17.

Sherwin, F. Intense Ice Age Volcanism Fits Biblical Model. *Creation Science Update.* Posted on ICR.org March 24, 2022.

Hebert, J. Volcanoes, Geoengineering, and the Post-Flood Ice Age. *Creation Science Update.* Posted on ICR.org April 2, 2020.

Hebert, J. The Tonga Volcano Eruption and the Ice Age. *Creation Science Update*. Posted on ICR.org February 14, 2022.

Clarey, T. 2019. Subduction Was Essential for the Ice Age. *Acts & Facts*. 48 (3): 9.

Clarey, T. Predicting Volcanic Eruptions Using Muography. *Creation Science Update*. Posted on ICR.org May 3, 2020.

Clarey, T. Volcanic Ash Turns to Stone in Months. *Creation Science Update*. Posted on ICR.org March 30, 2020.

Clarey, T. Massive Releases of CO_2 from Volcanism Rival Humans. *Creation Science Update*. Posted on ICR.org April 22, 2020.

Clarey, T. Express-Lane Magma Indicates Young Earth. *Creation Science Update*. Posted on ICR.org September 13, 2013.

Thomas, B. 2021. Haleakala National Park: One of Many Young-Looking Volcanoes. *Acts & Facts*. 50 (12): 14–17.

Clarey, T. 2017. Minuscule Erosion Points to Hawaii's Youth. *Acts & Facts*. 46 (1): 9.

Clarey, T. and J. J. S. Johnson. 2022. Crater Lake National Park: Serene Beauty After Volcanic History. *Acts & Facts*. 51 (1): 14–17.

Thomas, B. Discovery: Volcanoes on Venus. *Creation Science Update*. Posted on ICR.org July 13, 2015.

Thomas, B. Despite 'Magma Ocean' Discovery, Io's Volcanic Heat Remains a Mystery. *Creation Science Update*. Posted on ICR.org May 23, 2011.

Hebert, J. and T. Clarey. 2020. Catastrophic Plate Tectonics and the Flood. In *Creation Basics & Beyond An In-Depth Look at Science, Origins, and Evolution*, 2nd ed. Dallas, TX: Institute for Creation Research.

ABOUT THE INSTITUTE FOR CREATION RESEARCH

At the Institute for Creation Research, we want you to know God's Word can be trusted with everything it speaks about—from how and why we were made, to how the universe was formed, to how we can know Jesus Christ and receive all He has planned for us.

That's why ICR scientists have spent more than 50 years researching scientific evidence that refutes evolutionary philosophy and confirms the Bible's account of a recent and special creation. We regularly receive testimonies from around the world about how ICR's cutting-edge work has impacted thousands of people with Christ's creation truth.

HOW CAN ICR HELP YOU?

You'll find faith-building science articles in *Acts & Facts*, our bimonthly science news magazine, and spiritual insight and encouragement from *Days of Praise*, our quarterly devotional booklet. Sign up for FREE at **ICR.org/subscriptions**.

Our radio programs, podcasts, online videos, and wide range of social media offerings will keep you up-to-date on the latest creation news and announcements. Get connected at **ICR.org**.

We offer creation science books, DVDs, and other resources for every age and stage at **ICR.org/store**.

Learn how you can attend or host a biblical creation event at **ICR.org/events**.

Discover how science confirms the Bible at our Dallas museum, the ICR Discovery Center. Plan your visit at **ICRdiscoverycenter.org**.

ICR
INSTITUTE
FOR CREATION
RESEARCH

P. O. Box 59029
Dallas, TX 75229
800.337.0375
ICR.org

ICR'S MISSION STATEMENT
ICR EXISTS TO SUPPORT
THE LOCAL CHURCH THROUGH...

WORSHIP

- Glorify Jesus Christ by emphasizing in all ICR resources the credit He is due as Creator.

- Oppose the deification of nature by exposing Darwinian selectionism as an idolatrous worldview.

EDIFICATION

- Help pastors lead, feed, and defend their flocks by providing scientific responses to secular attacks on the authority and authenticity of God's Word.

- Change Christians' view of biology by constructing an organism-focused theory of biological design that highlights Jesus' work as Creator.

EVANGELISM

- Defend the gospel by showing how natural processes cannot explain the miracles in the Bible.

- Counter objections to the gospel by equipping believers with Scripture-affirming science.

RESOURCES FROM ICR

ICR's Creation Collection provides a biblical understanding of scientific topics. Written by Ph.D. scientists, each book focuses on a specific area of research that demonstrates both the unreliability of the evolutionary narrative and the infallibility of Scripture. Watch for upcoming books!

Find out more about these books and other resources at **ICR.org/store.**